SCIENTIFIC AMERICAN EXPLORES BIG IDEAS

The Science of Sleep

The Editors of *Scientific American*

SCIENTIFIC AMERICAN EDUCATIONAL PUBLISHING

New York

Published in 2024 by Scientific American Educational Publishing
in association with **The Rosen Publishing Group**
2544 Clinton Street, Buffalo NY 14224

Contains material from Scientific American®, a division of Springer Nature America, Inc., reprinted by permission, as well as original material from The Rosen Publishing Group®.

First Edition

Scientific American
Lisa Pallatroni: Project Editor

Rosen Publishing
Danielle Haynes: Compiling Editor
Michael Moy: Senior Graphic Designer

Cataloging-in-Publication Data
Names: Scientific American, Inc.
Title: The science of sleep / edited by the Scientific American Editors.
Description: First Edition. | New York : Scientific American Educational Publishing, 2024. | Series: Scientific American explores big ideas | Includes bibliographical references and index.
Identifiers: ISBN 9781725350489 (pbk.) | ISBN 9781725350496 (library bound)| ISBN 9781725350502 (ebook)
Subjects: LCSH: Sleep–Juvenile literature. | Sleep–Physiological aspects–Juvenile literature.
Classification: LCC QP425.S356 2024 | DDC 612.8'21–dc23

Manufactured in the United States of America
Websites listed were live at the time of publication.

Cover: Gorodenkoff/Shutterstock.com

CPSIA Compliance Information: Batch # CWSA24.
For Further Information contact Rosen Publishing at 1-800-237-9932.

CONTENTS

INTRODUCTION

Experts and doctors like to remind us that humans spend about one-third of our lives sleeping. That's assuming you get the recommended eight hours of shut eye. Spending that much time asleep, you better make sure it counts!

Sleep is critical to making sure you're awake and active each day. It's also one of the single-biggest factors in your health—it boosts your immune system, helps you recover from illness, and can even ward off physical and mental illness. For many people, though, it's not so easy to just go to bed, close your eyes, and fall asleep. Some people have troubles falling and staying asleep. Why do some people struggle so much and what are some ways to fix the problem beyond just counting sheep?

The articles in this book will explain the science behind why we sleep and how we can improve our quality of sleep. Section 1, "Why We Sleep," delves deeper into how sleep restores the human body and mind. One technique by inventor Thomas Edison even purports to boost creativity! Section 2, "The Secret to Better Sleep," explores scientific recommendations for getting better, more restful sleep. Are sleep drugs the answer, or are we better off simply seeking therapy? Section 3, "Poor Sleep and Health" is all about ways in which sleep—or lack thereof—can impact your health. Doctors have discovered links between rest and everything from diabetes and obesity to schizophrenia and dementia. Section 4, "The Importance of Good Rest for Kids and Teens," explores the different needs of young people and the link between sleep and getting a good education. Meanwhile, Section 5, "Snoozing Animals," shows how members of the animal kingdom get some Zs—some for months at a time—and even dream.

Section 1: Why We Sleep

1.1 Why Do We Sleep?
By Teodora Stoica

1.2 Spark Creativity with Thomas Edison's Napping Technique
By Bret Stetka

1.3 Deep Sleep Gives Your Brain a Deep Clean
By Simon Makin

1.4 New Hypothesis Explains Why We Sleep
By Giulio Tononi and Chaira Cirelli

1.5 Spiky 'Sleep Spindles' Linked to Acts of Learning
By Rebecca Sohn

1.6 What a Long-Ago Epidemic Teaches us About Sleep
By Christof Koch

Why Do We Sleep?

By Teodora Stoica

Try to explain to an alien why we sleep. Give it your best shot.

"Well, we get tired. And our brain needs to rest."

"I see. So, you find another way to defend yourself during rest?"

"Well ... no. Our body is paralyzed, and we lose consciousness."

There is an awkward pause. The alien tilts its head, feigning understanding.

"But! We sometimes dream!"

"Dream?"

Blindly, unaware of how ridiculous you sound, you continue with unprecedented speed and cadence: *"Yes! Dreams are fantastical stories projected from the mind* into *the mind, sometimes mixed with things that have* already *happened!"*

You catch your breath, and smile idiotically.

"And ... this helps with survival?"

"Well ... no. Sometimes the content confuses us in waking life," you suddenly realize.

The alien blinks silently a few times, furrowing his non-eyebrows:

"Let me see if I understand. Your species spends one third of their lives paralyzed watching fantasy movies?"

It is reasonable to assume the human race is doomed at this uncomfortable juncture in the conversation, and aliens will plan their stealthy attack during this incapacitated, seemingly futile stage of our existence.

But sleep isn't futile. In fact, theories on sleep assert its role in reenergizing the body's cells, clearing waste from the brain, and supporting learning and memory. It can even regulate mood, appetite and libido. Yet what precisely happens in our head when it hits the pillow? E.T., you got some learnin' to do.

Reflections on why we sleep have existed since the beginning of recorded history. Some 2,000 years ago, Leucippus, the first "atomist," described sleep as something that happened to the body when "the excretion of fine-textured atoms exceeds the accretion of psychic warmth." Fanciful.

Another famed ancient, Lucretius, thought sleep is caused by the spirit being attacked and weakened into a state of senescence by mysterious air particles. Absurd. Lastly, Aristotle distinguished sleep and waking as diametrically opposed phenomena characterized by the absence or presence of perception. On the money. He stated: "Sleep is induced by the exhalations of ingested foods which thicken and heat the blood, rising to the brain where they are cooled before coalescing in the heart."

Well at least he had the brain part right. Indeed, electroencephalographic (EEG) data measuring brain waves have categorized sleep into two states: non-REM (or NREM, further divided into stages S1, S2, S3 and S4) and REM, which repeat every 90 minutes several times during the night.

When overcome by drowsiness but still awake, brain waves become slower, increase in height (amplitude) and become more synchronous (alpha waves). The first official stage of sleep serves as transition between awake and sleep, and is characterized by theta waves, which are even slower in number (frequency) and greater in amplitude than alpha waves. Many sleep-deprived individuals experience microsleeps (frighteningly, while driving!), which are second-long temporary sleep episodes where theta waves replace alpha wave activity. Most people are not even aware they were asleep.

During stage 2 sleep, theta wave activity continues, interspersed with oscillations between the thalamus, a region of the brain regulating sleep, and the rest of the cortex. Stages 1 and 2 are relatively "light" stages of sleep, where waking up is easy and the sleeper might not recognize they were asleep at all.

Onto stage 3 and 4, characterized by delta waves, the slowest and highest amplitude brain waves, the most unlike waking brain waves. Since delta sleep is the deepest sleep, it is the most difficult stage in which to wake sleepers, and when they are awakened, they are usually sleepy and disoriented. Interestingly, delta sleep is when sleepwalking and sleep talking is most likely to occur.

During NREM sleep, sensory information continues to be transferred and encoded in a similar way to the brain when awake—especially at the level of sensory cortices. As sleep progresses, reactivity and awareness decrease. Perceptual activity becomes progressively autonomous. The mind becomes focused inwards and becomes less constrained by input from the external environment.

This pattern is visualized in analyses examining resting-state networks, networks of brain regions activated synchronously. The data show that two networks remain active similarly to when awake: the default mode network and sensory motor network. The default mode network is activated during self-generated thought (daydreaming), while the sensory network is activated during stimulus perception. Connectivity to the central executive network, a network active during higher-order processing such as inhibition, is reduced, letting thoughts roam fancy free and creating illogical scenarios where dragons mothered by a platinum beauty queen roam the planet—for example.

Finally, the one Freud was all excited about: REM. This stage gets its name from the darting eye movements that accompany it (Rapid Eye Movement). Interestingly, it is also characterized by a sudden and dramatic loss of muscle tone—effectively paralysis. Importantly, it is neurologically identical to waking (dyssynchronous brain waves) and further characterized by variable changes in blood pressure, heart rate and overall breathing rate. And of course, dreams.

Plato was the first to muse psychologically about dreams, stating: "In all of us, even the most highly respectable, there is a lawless wild beast nature, which peers out in sleep." This presages Freud's wish-fulfillment theory of dreaming yet doesn't neurologically explain that thing Martin Luther King, Jr., had in 1963.

Firstly, dreams are *not* hallucinations. While they share some characteristics in common with clinical disorders such as schizophrenia, hallucinations are defined by whole-brain activation. They are veridical movies projected onto reality that the individual remembers well. Dreams, in contrast, are characterized by primary and secondary sensory cortices activation, coupled with disrupted recruitment of the frontal cortex, gate keeper of important higher-order functions like inhibition. If the dream is imbibed in high emotional content, the limbic system is also implicated.

Critically, lack of frontal function opens the floodgates to reduced self-awareness and bizarre dream contents, while restricting any incoming information and creating a tight self-contained dream loop. This is perhaps most poetically described by Heraclitus, who commented on the subjective nature of dreams in his memorable phrase, "for those who are awake there is a single, common universe, whereas in sleep each person turns away into his own private universe."

A note on lucid dreaming—being aware that you're in a dream. A few studies have found evidence for increased activation of the prefrontal cortex during lucid dreaming, compared to normal REM sleep. The area's typical *de*activation is responsible for the socially inappropriate behaviors and inability to reason logically that is sometimes experienced during dreams, if they are remembered upon waking.

In case you don't remember your dreams, you can always Google them. Seriously. The Google artificial neural network is like a computer brain, inspired by the central nervous system of animals. When given free rein to create images out of white noise, it outputs beautiful, intricate, Escher-esque images, what Google engineers are terming "AI Dreams." The neural network's dreams are being used to understand how a computer can make sense out of nonsense—or learn.

We've arrived at E.T.'s original inquiry: "But *why* do we sleep?" The easiest answer is, we don't yet know, but theories abound. One theory has to do with sleep playing an important role in memory consolidation, that is, a memory's ability to stick. Research suggests REM sleep strengthens weakened neural circuits holding past memories to prepare them for future cognitive processing, in addition to filtering out unnecessary memories. A study by Poe, et. al (2014) articulates this: "The evidence points to an overall function of embossing of synaptic circuits to add definition to and integrate memories, distinguishing them from the noise introduced to synaptic weights throughout wakefulness."

Termed the "synaptic homeostasis hypothesis," it suggests that during sleep there is widespread weakening of connections (synapses) throughout the brain. This is thought to counterbalance the overall strengthening of connections that occurs during learning when we are awake. By pruning away excess connections, sleep effectively "cleans the slate" so we can learn again the next day. Interfering with this scaling down process can, in some cases, lead to more intense (and perhaps unwanted) memories.

Another theory, since E.T. doesn't seem impressed, is that sleep is restorative: it flushes out toxins. During sleep, the space between brain cells increases, allowing toxic proteins to be flushed out. It's possible that by removing these toxins from the brain, sleep may stave off neurodegenerative diseases like Alzheimer's. Indeed, in an October 2013 issue of the journal *Science*, researchers published the results of a study indicating that the brain utilizes sleep to flush out toxins. This waste removal system, they suggest, is one of the major reasons why we sleep.

While we don't have a precise overarching definition of sleep's function, it impacts different physiological and psychological purposes including cleaning up brain toxins and consolidating information into memory, so it is possible that each of these theories can be used to explain why we sleep; we just haven't worked it out. Give us a break, E.T., we're asleep most of the time.

The views expressed are those of the author(s) and are not necessarily those of Scientific American.

About the Author

Teodora Stoica is a Ph.D. candidate in the Translational Neuroscience Department at the University of Louisville. She writes about science on her personal blog CuriousCortex and is a member of the Communications Committee at the Organization for Human Brain Mapping. You can follow her on Twitter @CuriousCortex.

Spark Creativity with Thomas Edison's Napping Technique

By Bret Stetka

Thomas Edison was famously opposed to sleeping. In an 1889 interview published in *Scientific American*, the ever energetic inventor of the lightbulb claimed he never slept more than four hours a night. Sleep was, he thought, a waste of time.

Yet Edison may have relied on slumber to spur his creativity. The inventor is said to have napped while holding a ball in each hand, presuming that, as he fell asleep, the orbs would fall to the floor and wake him. This way he could remember the sorts of thoughts that come to us as we are nodding off, which we often do not recall.

Sleep researchers now suggest that Edison might have been on to something. A study published recently in *Science Advances* reports that we have a brief period of creativity and insight in the semilucid state that occurs just as we begin to drift into sleep, a sleep phase called N1, or nonrapid-eye-movement sleep stage 1. The findings imply that if we can harness that liminal haze between sleep and wakefulness–known as a hypnagogic state–we might recall our bright ideas more easily.

Inspired by Edison, Delphine Oudiette of the Paris Brain Institute and her colleagues presented 103 participants with mathematical problems that had a hidden rule that allowed them to be solved much faster. The 16 people who cracked the clue right away were then excluded from the study. The rest were given a 20-minute break period and asked to relax in a reclined position while holding a drinking glass in their right hand. If it fell, they were then asked to report what they had been thinking prior to letting go.

Throughout the break, subjects underwent polysomnography, a technology that monitors brain, eye and muscle activity to assess a person's state of wakefulness. This helped to determine which subjects were awake rather than in N1 or if they were in N2–the next, slightly deeper phase of our sleep.

After the break, the study subjects were presented with the math problems again. Those who had dozed into N1 were nearly three times more likely to crack the hidden rule as others who had stayed awake throughout the experiment–and nearly six times more likely to do so as people who had slipped into N2. This "eureka moment," as the authors call it, did not occur immediately. Rather it happened after many subsequent attempts to solve the math problem, which is consistent with previous research on insight and sleep.

It's less clear that Edison's technique of dropping objects to ward off deeper sleep works. Of the 63 subjects who dropped the glass as they drowsed, 26 did so after they had already passed through N1 sleep. Still, the findings suggest that we do have a creative window just before falling asleep.

Oudiette says that, like Edison, her personal experience with sleep inspired the study. "I've always had a lot of hypnagogic experiences, dreamlike experiences that have fascinated me for a long time," she says. "I was quite surprised that almost no scientists have studied this period in the past two decades."

A study published in 2018 found that a brief period of "awake quiescence," or quiet resting, increased the odds of discovering the same mathematical rule used in Oudiette's experiment. And psychologist Penny Lewis of Cardiff University in Wales suggests, that both rapid-eye-movement (REM) sleep–the phase in which our eyes dart back and forth and most dreams occur–and non-REM sleep work together to encourage problem-solving.

Yet for the most part, Oudiette is not aware of any other research specifically looking at the influence of sleep onset on creativity. She does, however, point to plenty of historical examples of this phenomenon.

"Alexander the Great and [Albert] Einstein potentially used Edison's technique, or so the legend goes," she says. "And some of the dreams that have inspired great discoveries could be hypnagogic experiences rather than night dreams. One famous example is the chemist August Kekulé finding the ring structure of benzene after seeing a snake biting its own tail in a 'half-sleep' period when he was up working late." Surrealist painter Salvador Dalí also used a variation of Edison's method: he held a key over a metal plate as he went to sleep, which clanged to wake him as he dropped it, supposedly inspiring his artistic imagery.

"This study gives us simultaneous insight into consciousness and creativity," says Adam Haar Horowitz of the M.I.T. Media Lab, who has devised technology to interact with hypnagogic states but did not collaborate with Oudiette's team. "Importantly," he adds, "it's the kind of study that you can go ahead and try at home yourself. Grab a metal object, lie down, focus hard on a creative problem, and see what sort of eureka moments you can encounter."

For University of California, Santa Barbara, psychologist Jonathan Schooler, who also was not involved with the project, the study does not necessarily prove that just anyone will be able to mine their creativity during this early phase of somnolence. As he points out, "residing in the 'sweet zone' might have also simply refreshed the study participants, making it easier for them to solve the problem later." But Schooler acknowledges there may be something very solid in the study's findings. "The new results suggest there is a creative sleep sweet spot during which individuals are asleep enough to access otherwise inaccessible elements but not so far gone the material is lost," he says.

Despite its reputation as the brain's period of "shutting off," sleep is, neurologically speaking, an incredibly active process. Brain cells fire by the billions, help to reactivate and store memories, and, it seems, allow us to conjure our mental creations.

Oudiette hopes not only to confirm her findings in future research but also to determine if focusing on our hypnagogic state might help solve real-world tasks and problems by harnessing the creative potential of that liminal period between sleep and wakefulness. Additionally, she and her group are considering the potential of brain-computer interfaces to precisely identify brain-wave patterns associated with the onset of sleep, allowing the precise identification of when people should be woken up during their moments of putative insight.

"We could even teach people how to reach this creative state at will," Oudiette envisions. "Imagine playing sounds when people are reaching the right state and other sounds when they are going too far into sleep. Such a method could teach them how to recognize the creative state and how to reach it."

About the Author

Bret Stetka is a writer based in New York City and editorial director of Medscape Neurology (a subsidiary of WebMD). His work has appeared in Wired, *NPR and the* Atlantic. *He graduated from the University of Virginia School of Medicine in 2005.*

Deep Sleep Gives Your Brain a Deep Clean

By Simon Makin

Why sleep has restorative—or damaging—effects on cognition and brain health has been an enduring mystery in biology. Researchers think cerebrospinal fluid (CSF) may flush toxic waste out, "cleaning" the brain, and studies have shown that garbage clearance is hugely improved during sleep. Scientists were not sure exactly how all this works, however, or why it should be so enhanced during sleep.

One aspect of sleep that is well understood is how the slow electrical oscillations (or "slow waves") that characterize deep, non-REM sleep contribute to memory consolidation, the process whereby new memories are transferred into long-term storage. A 2019 study, from a team led by neuroscientist Laura Lewis of Boston University, gives insight into what drives CSF flow through the brain, suggesting that the same slow waves that coordinate memory consolidation drive oscillations in blood flow and CSF in the brain.

The work has implications for understanding the relations between sleep disturbance and psychiatric and neurodegenerative conditions, and it may even point to new approaches to diagnosis and treatment. "We've discovered there are really large waves of CSF that appear in the brain only during sleep," Lewis says. "This effect is really striking, and we're also interested in what it means for maintaining brain health, especially in disorders such as Alzheimer's disease."

In the study, published in *Science*, the team set out to investigate how the dynamics of CSF flow change during sleep and how such changes might relate to alterations in brain blood flow and electrical activity. "We know sleep is really important for brain health, and waste clearance is probably a key reason why. What was less clear is: Why is this changed during sleep?" Lewis says. "That led us to ask what was happening in the CSF."

The researchers used electroencephalography (EEG) to monitor the brain waves of 13 sleeping healthy adults, while also using a cutting-edge, "accelerated" functional magnetic resonance imaging technique to capture faster changes than standard fMRI can manage. That allowed them to measure both blood-oxygenation changes (which indicate blood flowing to electrically active, oxygen-hungry regions) and CSF flows. The latter was possible only because of a flaw in this method that means any newly arriving fluid (not just oxygenated blood) lights up in the image. "We realized we could take advantage of this to measure CSF flow at the same time as blood oxygenation," Lewis says. "That was critical because it turns out these things are coupled to each other in a way we never would have seen if we didn't measure blood, CSF and electrical activity simultaneously."

What the team found was that the slow waves seen in non-REM sleep occur in lockstep with changes in both blood flow and CSF. Just because things occur together doesn't necessarily mean one causes the other, but the team also built a computer model incorporating information about the physics linking these processes, which predicted that slow waves would have just these kinds of effects on blood and CSF. What seems to be happening is that as brain activity alters blood flow, the volume of blood in the brain decreases, and because the brain is a closed vessel, CSF flows in to fill the space. "It's very convincing," says neurologist Maiken Nedergaard of the University of Rochester, who was not involved with the research. "It also really makes sense: electrical activity drives blood flow changes that then drive CSF changes."

The team measured this CSF inflow going into the fourth ventricle, one of four fluid-filled cavities involved in producing CSF (by filtering blood plasma) and circulating it around the brain. As CSF usually flows out of the fourth ventricle, this suggests a "pulsatile" flow, like a wave. This pushes CSF around the ventricles and into spaces between membranes surrounding the brain and spinal cord, called the meninges, where it mixes with "interstitial fluid" within the brain to carry away toxic waste products.

Because slow waves are important for memory consolidation, this links two disparate functions of sleep. "What's exciting about this is it's combining features of brain function that people don't normally think of as connected," Nedergaard says. It isn't obvious things had to be this way, Lewis says, but it may represent an example of nature being efficient. "It's a matter of nature not dividing tasks between higher level and lower level, like how you run a company, where you have a boss making decisions and cleaning people coming in," Nedergaard says. "In biology, it's everybody contributing, as it makes more sense."

The findings have implications for neurodegenerative diseases, which might be caused by buildup of toxic proteins in the brain, such as beta-amyloid in Alzheimer's disease. Previous research has shown that beta-amyloid is cleared more efficiently during sleep, which is often disrupted in patients. Disturbances in slow-wave sleep also often accompany aging, which may be linked to cognitive decline. "We know that people with Alzheimer's have fewer slow waves, so we may find they also have fewer CSF waves," Lewis says. "We have to do these studies now in older adults and patient populations to understand what this might mean for those disorders." Sleep disturbance is also a feature of many psychiatric disorders, from depression to schizophrenia. "Different electrical signatures of sleep are disrupted in different psychiatric conditions," she says. "So this will be very interesting to follow up on in a multitude of disorders."

The team has talked about the possibility of modulating brain activity to try to nail down whether electrical oscillations truly do cause the changes they observed in CSF flow. "It would be great to find the right collaborator and do a study in mice where we manipulate neural activity, then watch the downstream consequences," Lewis says. "We're also thinking about ways to safely and noninvasively manipulate neural oscillations in humans." It may be possible to use electromagnetic stimulation to influence brain waves as a treatment for brain disorders. Researchers have seen encouraging results of this approach in mice, and these findings may help explain why.

Another potential application may come from assessing whether changes in CSF flows can serve as a diagnostic marker for some of these conditions. "It gives us a ton of interesting new biology to explore and understand since it seems like things the brain is doing during sleep are related to each other in surprising ways," Lewis says. "Maybe the most important take-home message is that sleep is a serious thing," Nedergaard says. "You really need to sleep to keep a healthy brain because it links electrical activity to a practical housekeeping function."

About the Author

Simon Makin is a freelance science journalist based in the U.K. His work has appeared in New Scientist, *the* Economist, Scientific American *and* Nature, *among others. He covers the life sciences and specializes in neuroscience, psychology and mental health. Follow Makin on Twitter @SimonMakin.*

New Hypothesis Explains Why We Sleep

By Giulio Tononi, Chiara Cirelli

Every night, while we lie asleep, blind, dumb and almost paralyzed, our brains are hard at work. Neurons in the sleeping brain fire nearly as often as they do in a waking state, and they consume almost as much energy. What is the point of this unceasing activity at a time when we are supposedly resting? Why does the conscious mind disconnect so completely from the external environment while the brain keeps nattering on?

The brain's activity during rest likely serves some essential function. The evidence for this importance starts with sleep's ubiquity. All animals apparently sleep even though being unconscious and unresponsive greatly raises the risk of becoming another creature's lunch. Birds do it, bees do it, iguanas and cockroaches do it, even fruit flies do it, as we and others demonstrated more than a decade ago.

Furthermore, evolution has devised a few extraordinary adaptations to accommodate sleep: dolphins and some other marine mammals that must surface often to breathe, for example, sleep by alternately switching off one hemisphere of their brain while the other remains in a waking state.

Like many scientists and nonscientists, the two of us have long wondered what benefit sleep provides that makes it so crucial to living creatures. More than 20 years ago, when we worked together at the Sant'Anna School of Advanced Studies in Pisa, Italy, we began to suspect that the brain's activity during slumber may somehow restore to a baseline state the billions of neural connections that get modified every day by the events of waking life. Sleep, in this telling, would preserve the ability of the brain's circuitry to form new memories continually over the course of an individual's lifetime without becoming oversaturated or obliterating older memories.

We also have an idea of why awareness of the external environment must be shut off during sleep. It seems to us that conscious experience of the here and now has to be interrupted for the brain to gain the chance to integrate new and old memories; sleep provides that respite.

Our hypothesis is somewhat controversial among our fellow neuroscientists who study sleep's role in learning and memory because we suggest that the return to baseline results from a weakening of the links among the neurons that fire during sleep. Conventional wisdom holds, instead, that brain activity during sleep *strengthens* the neural connections involved in storing newly formed memories. Yet years of research with organisms ranging from flies to people lend support to our notions.

School of Nod

Scientists first proposed the idea that sleep is important to memory nearly a century ago, and plenty of experiments since then have shown that after a night of sleep, and sometimes just a nap, newly formed memories "stick" better than they would if one had spent the same amount of time awake. This pattern holds for declarative memories, such as lists of words and associations between pictures and places, as well as for procedural memories, which underlie perceptual and motor skills, such as playing a musical instrument.

The evidence that sleep benefits memory led scientists to look for signs that the brain rehashes newly learned material at night. They found them: studies performed over the past 20 years, first in rodents and then in humans, show that patterns of neural activity during sleep sometimes do resemble those recorded while subjects are awake. For example, when a rat learns to navigate a maze, certain neurons in a part of the brain called the hippocampus fire in specific sequences. During subsequent sleep, rats "replay" these sequences more often than predicted by chance.

Because of such findings, many researchers came to assume that sleep "replay" consolidates memories by further reinforcing synapses–the contact points between neurons–that have been strengthened when an individual is awake. The idea is that, as linked neurons fire repeatedly, the synapses connecting them more readily convey signals from one neuron to another, helping neuronal circuits to encode memories in the brain. This process of selective strengthening is known as synaptic potentiation, and it is the favored mechanism by which the brain is thought to accomplish learning and remembering.

Yet while replay and potentiation are known to occur during waking activities, scientists have so far found no direct evidence that the synapses in replayed circuits get strengthened during sleep. This lack of evidence hardly surprises us. It is consistent with our suspicion that while the sleeper lies unaware, all that brain activity—the "replay" as well as other, seemingly random firings—might actually be *weakening* neural connections, not strengthening them.

The Price of Plasticity

There are many good reasons to propose that synapses must become weakened as well as strengthened for the brain to function properly. For one thing, strong synapses consume more energy than weak ones, and the brain does not have infinite stores of energy. In humans the brain accounts for almost 20 percent of the body's energy budget—more than any other organ by weight—and at least two thirds of that portion goes to supporting synaptic activity. Building and bolstering synapses is also a major source of cellular stress, requiring cells to synthesize and deliver components ranging from mitochondria (the cell's power plants), to synaptic vesicles (which ferry signaling molecules), to various proteins and lipids that are needed for communication across synapses.

It seems clear to us that this strain on resources is unsustainable. The brain cannot go on strengthening and maintaining revved-up synapses both day and night for the whole of an individual's lifetime. We do not doubt that learning occurs mainly through synaptic potentiation. We simply doubt that strengthening continues to happen during sleep.

In contrast, synaptic weakening during sleep would restore brain circuitry to a baseline level of strength, thereby avoiding excessive energy consumption and cellular stress. We refer to this baseline-restoring function of sleep as preserving synaptic homeostasis, and we call our overall hypothesis about the role of sleep the synaptic homeostasis hypothesis, or SHY. In principle, SHY explains the essential, universal purpose of sleep for all organisms that do it: sleep restores the brain to a state where it can learn and adapt when we are awake. The risk we take by becoming disconnected from the environment for hours at a time is the price we pay for this neural recalibration. Most generally, sleep is the price we pay for the brain's plasticity–its ability to modify its wiring in response to experience.

But how does SHY explain sleep's salutary effects on learning and memory? How can weakened synapses improve the overall retention of skills and facts? Consider that, over the course of a typical day, almost everything you experience leaves a neural trace in the brain and that the significant events, like meeting a new person or learning a piece of music on the guitar, make up just a trifling portion of that neural encoding. To improve memory, the sleeping brain must somehow distinguish the "noise" of irrelevant information from the "signal" of significant happenings.

We suggest that in sleep, the spontaneous firing of neurons in the brain activates many different circuits in many different combinations, encompassing both new memory traces and old networks of learned associations. (You get a glimpse of this neural free-for-all in dreams.) The spontaneous activity lets the brain try out which new memories fit better with stored memories of proved significance and weakens those synapses that do not fit well in the grand scheme of memory. We and other investigators are exploring possible mechanisms by which brain activity could selectively weaken synapses that encode the "noise" while preserving those that correspond to the "signal."

While the brain tries out these imaginary scenarios and enacts weakening where appropriate, we had best be unaware of the surrounding environment and be incapable of acting in it; that is, we had best be asleep. Likewise, restoring synaptic homeostasis should not take place while we are awake because the events of the day would dominate the process, giving salience to them rather than to all the knowledge the brain has accumulated over a lifetime. The profound disconnection of sleep frees our brain from the tyranny of the present, creating an ideal circumstance for integrating and consolidating memories.

A Weak Connection

Our proposal that the brain uses neuronal firing during sleep to weaken rather than strengthen synapses is supported in part by close analyses of data from a standard workhorse of sleep research: the electroencephalogram, or EEG. EEGs record patterns of electrical activity in the cerebral cortex via electrodes attached to the scalp. Decades ago EEG recordings of the sleeping brain revealed two main categories of sleep, called rapid eye movement (REM) and non-REM (NREM), that alternate throughout the night. Each has distinctive brain-wave patterns. In addition to the jittering of eyeballs underneath closed lids that gives REM sleep its name, that stage is dominated by relatively fast oscillations–quick ups and downs in the curves of the EEG readout, resembling EEG recordings of the waking state. In contrast, slow oscillations–with frequencies of about one cycle per second–are the most prominent feature of NREM sleep.

A decade ago the late Mircea Steriade of Laval University in Quebec discovered that the slow oscillations of NREM sleep arise when groups of neurons fire together for a little while (so-called on periods), then fall silent for about a fraction of a second (off periods) and then resume their synchronized firing. This was one of the fundamental discoveries in sleep research. Since then, scientists have also discovered that in birds and mammals, the slow waves are large if preceded by a long period of wakefulness and become smaller as sleep goes on.

We reasoned that if synapses are strong, neurons will synchronize their firing more, producing larger slow waves. If synapses are weak, neurons will be less synchronized and the resulting slow waves will be smaller. Results of computer simulations and experiments in humans and animals led us to conclude that the big, steep slow waves early in the night indicate that synapses have been strengthened by prior wakefulness, whereas the small, shallow slow waves early in the morning indicate that synapses have become weaker during sleep.

Direct support for the idea that synapses become weaker during sleep, and may even be pruned away, comes from studies in animals. In fruit flies, for instance, we find that sleep reverses a progressive increase in the number and size of synapses that occurs during the day, especially when the flies are exposed to stimulating environments. Synaptic spines are specialized protrusions on a neuron's signal-detecting arm. When fruit flies spend the day interacting with other flies, neurons throughout their brain sprout more synaptic spines by evening than were present in the morning.

Just as remarkably, the number of spines goes back to the baseline level by the following morning if—and only if—the flies are allowed to sleep. We saw a similar phenomenon in the cerebral cortex of adolescent mice: the number of synaptic spines tended to rise when the animals were awake and to fall when they slept. In adult rodents, the upshot is the same, although it is not the number of synaptic spines that changes with wakefulness and sleep but rather the abundance of certain spine molecules, known as AMPA receptors, that determine the strength of a synapse. When we monitored these AMPA receptors, we found that their number per synapse increases after wakefulness and decreases after sleep. More receptors make for stronger synapses; fewer mean the synapses have weakened.

Synaptic strength can be gauged directly by using an electrical probe to stimulate neural fibers in the cortex. The neurons respond with an induced electrical discharge that is larger when synapses are strong and smaller when the connections are weak. We showed that in rats, stimulated neurons fire more strongly after a few hours of wakefulness and less strongly after sleep. Marcello Massimini of the University of Milan in Italy and Reto Huber, now at the University of Zurich, performed a similar experiment in humans. Instead of an electrical probe, they turned to transcranial magnetic stimulation—a short magnetic pulse applied to the scalp—to stimulate the underlying neurons. They then recorded the strength of the cortical responses with high-density EEG. The results were clear: the longer a subject was awake, the larger the EEG responses. It took a night of sleep for cortical responses to return to the baseline.

Less Is More

The common conclusion of these experiments, which we performed over two decades, is that spontaneous cortical activity in sleep does indeed weaken the synaptic connections in neural circuits, whether by damping their ability to send electrical impulses or by erasing them outright.

This process, which we call down selection, would ensure the survival of the circuits that are "fittest," either because they were activated strongly and consistently during wakefulness (say, by playing the right notes on a guitar while trying to master a new piece) or because they were better integrated with previous, older memories (as would be the case for a new word encountered in a known language). Meanwhile synapses in circuits that were only mildly enhanced during wakefulness (such as fumbled notes on the guitar) or that fit less with old memories (such as a new word presented in an unknown language) would be depressed.

Down selection would ensure that insignificant events would leave no lasting trace in our neural circuitry, whereas memories of note would be preserved. As an additional bonus, down selection would also make room for another cycle of synaptic strengthening during wakefulness. Indeed, some findings imply that among its many other benefits for learning and memory, sleep aids the subsequent acquisition of new memories (material encountered before the next bout of sleep). Quite a few studies have shown that after a night of sleep, you can learn new material much better than you can after having been awake all day. (Students, take note.)

Although we have no direct evidence for a mechanism that would produce selective weakening of activated synapses as yet, we have a notion of how synaptic weakening could occur. We suspect the slow waves of mammalian NREM sleep somehow play a role. In lab studies of rat brain tissue, nerve cells became less effective at passing signals to one another when stimulated in ways that mimic the synchronized on/off cycles of slow-wave sleep.

The chemistry of the brain also changes in NREM sleep in a way that could lead to synaptic weakening. In the awake individual, a concentrated soup of signaling chemicals, or neuromodulators–including acetylcholine, norepinephrine, dopamine, serotonin, histamine and hypocretin–bathe the brain and bias synapses toward strengthening when signals pass through them. During sleep–especially NREM sleep–the soup becomes much less concentrated. This diluted milieu of neuromodulators may bias the neural circuitry so that synapses become weakened, rather than strengthened, when signals flow across them. The process might also involve a substance called brain-derived neurotrophic factor (BDNF), which is known to promote synaptic strengthening and to be involved in memory acquisition. BDNF levels are high in neurons during wakefulness and minimal during sleep.

Local Sleep

Regardless of specific mechanisms and selective processes, the evidence is strong in several species that overall synaptic strength goes up during wakefulness and down during sleep: the core prediction of SHY. We can test SHY further by examining some of its intriguing corollaries.

For example, if the hypothesis is correct, then the more plasticity a part of the brain undergoes during wakefulness, the more that part should need to sleep. "Sleep need" can, in turn, be indicated by an increase in the size and duration of NREM slow waves. To explore this prediction, we asked human subjects to learn a novel task: how to reach a target on a computer screen while the cursor (controlled by a mouse) is systematically rotated. The part of the brain that engages in this kind of learning is the right parietal cortex. Sure enough, when our subjects slept, the slow waves over their right parietal cortex were larger, relative to waves from the same area on the night before learning occurred. These large waves did flatten out in the course of the night, as such oscillations do. But those large, localized waves at the start of the night tell us that particular part of the brain had been exhausted by the task we assigned.

Many other experiments by the two of us and others have since confirmed that learning, and more generally the activation of synapses in circuits, produces a local increase in sleep need. Recently we have even found that prolonged or intense use of certain circuits can make local groups of neurons "fall asleep" even though the rest of the brain (and the organism itself) remains awake. Thus, if a rat stays awake longer than usual, some cortical neurons show brief periods of silence that are basically indistinguishable from the off periods observed during slow-wave sleep. Meanwhile the rat is running around, its eyes open, tending to its business, as any awake rat would do.

This phenomenon is called local sleep, and it is attracting scrutiny from other investigators. Our latest studies indicate that localized off periods also occur in the brains of sleep-deprived humans and that those periods become more frequent after intense learning. It seems that when we have been awake for too long or have overexerted certain circuits, small chunks of the brain may take quick naps without giving notice. One wonders how many errors of judgment, silly mistakes, irritable responses and foul moods result from local sleep in the brains of exhausted people who believe they are fully awake and in complete control.

SHY also predicts that sleep is especially important in childhood and adolescence, times of concentrated learning and of intense synaptic remodeling, as many studies have shown. In youth, synapses are formed, strengthened and pruned at an explosive rate never approached in adulthood. It makes sense that down selection during sleep would be crucial to minimize the energy costs of this frenzied synaptic remodeling and to favor the survival of adaptive neural circuits in these stages of life. One can only wonder what happens when sleep is disrupted or insufficient during critical periods in development. Might the deficit corrupt the proper refinement of neural circuits? In that case, the effect of sleep loss would not merely be occasional forgetfulness or misjudgment but a lasting change in the way the brain is wired.

We look forward to testing SHY's predictions and exploring its implications further. For example, we hope to discover whether sleep deprivation during neural development leads to changes in the organization of brain circuitry. We would also like to learn more about the effect of sleep on deep-brain areas, such as the thalamus, cerebellum, hypothalamus and brain stem, and about the role of REM sleep in synaptic homeostasis. Perhaps we would then learn if sleep is indeed the price of waking plasticity, a price that every brain and every neuron must pay.

Referenced

Is Sleep Essential? Chiara Cirelli and Giulio Tononi in *PLOS Biology*, Vol. 6, No. 8, pages 1605–1611; August 2008.

The Memory Function of Sleep. Susanne Diekelmann and Jan Born in *Nature Reviews Neuroscience*, Vol. 11, No. 2, pages 114–126; February 2010.

Local Sleep in Awake Rats. Vladyslav V. Vyazovskiy, Umberto Olcese, Erin C. Hanlon, Yuval Nir, Chiara Cirelli and Giulio Tononi in *Nature*, Vol. 472, pages 443–447; April 28, 2011.

Sleep and Synaptic Homeostasis: Structural Evidence in Drosophila. Daniel Bushey, Giulio Tononi and Chiara Cirelli in *Science*, Vol. 332, pages 1576–1581; June 24, 2011.

About the Authors

Giulio Tononi and Chiara Cirelli are professors of psychiatry at the University of Wisconsin–Madison. Their research on the function of sleep is part of a larger investigation of human consciousness, the subject of Tononi's recent book Phi: A Voyage from the Brain to the Soul *(Pantheon, 2012).*

Spiky 'Sleep Spindles' Linked to Acts of Learning

By Rebecca Sohn

Distinctive bursts of sleeping-brain activity, known as sleep spindles, have long been generally associated with strengthening recently formed memories. But new research has managed to link such surges to specific acts of learning while awake.

These electrical flurries, which can be observed as sharp spikes on an electroencephalogram (EEG), tend to happen in early sleep stages when brain activity is otherwise low. A study published in *Current Biology* shows that sleep spindles appear prominently in particular brain areas that had been active in study participants earlier, while they were awake and learning an assigned task. Stronger spindles in these areas correlated with better recall after sleep.

"We were able to link, within [each] participant, exactly the brain areas used for learning to spindle activity during sleep," says University of Oxford cognitive neuroscientist Bernhard Staresina, senior author on the study. Staresina, Marit Petzka of the University of Birmingham in England and their colleagues devised a set of tasks they called the "memory arena," which required each participant to memorize a sequence of images appearing inside a circle. While the subjects did so, researchers measured their brain activity with an EEG, which uses electrodes placed on the head. Participants then took a two-hour nap, after which they memorized a new image set–but then had to re-create the original image sequence learned before sleeping.

During naps, the researchers recorded stronger sleep spindles in the specific brain areas that had been active during the pre-sleep-memorization task, and these areas differed for each participant. This suggested that the spindle pattern was not "hardwired" in default parts of the human brain; rather it was tied to an individual's thought patterns. The researchers also observed that participants who experienced stronger sleep spindles in brain areas used during memorization did a better job re-creating the images' positions after the nap.

Previous research suggested spindles trigger brain changes that start the process of strengthening and refining a memory, in part by controlling the influx of calcium to certain cells. The new study is the first to directly measure brain activity during learning to support such a link and the first to associate better memory recall with stronger spindles in areas that had been active during learning, the researchers say—although they note it does not indicate whether spindles themselves actively improved recall. Staresina says that future work should examine spindle and other activity in the hippocampus, the brain's memory hub.

The study is "a convincing manuscript that supports the role of spindles in memory," says Lourdes DelRosso, a sleep researcher and physician at the University of Washington and Seattle Children's Hospital, who was not involved in the work. She hopes scientists also investigate spindles in people with conditions that affect learning and attention, such as ADHD and dyslexia.

About the Author

Rebecca Sohn is a freelance health and science journalist based in New York City. She has written for outlets such as Slate, Spectrum, STAT *and* Live Science, *among others.*

What a Long-Ago Epidemic Teaches Us about Sleep

By Christof Koch

> Unity Kinkaid finds it harder and harder to stay awake. She now sleeps for almost 20 hours a day. She used to dream; to shift in her sleep, muttering and sighing, locked in half-remembered fantasies ... Now she lies unmoving, breath shallow and silent, lost to the world. Unity sleeps.
>
> –Neil Gaiman, *The Sandman: Preludes & Nocturnes*, 1991

In the above-cited comic, a necromancer imprisons Morpheus, an immortal who is master of sleep and dreams. Consequently, people are unable to stay awake, sleeping the days, months and years away, until Morpheus escapes and restores the natural order. The chilling real-life inspiration for this compelling serialized comic was an epidemic that swept the world between 1916 and 1926. A hundred years after the first cases of "sleepy sickness" came to the attention of the medical community, we still do not know what agents were responsible for this disease.

It emerged from the cold, wet plains and trenches of northern France and Belgium, the battlefields of World War I, shorn apart and denuded of trees by the endless barrage of explosions from artillery, mines and machine-gun fire, a landscape crowded with millions of young men living under atrocious and unsanitary conditions at close quarters. Indeed, one of the most famous victims of the disease may have been Adolf Hitler, who was wounded at the Battle of the Somme.

Two physicians from opposing sides, Jean-René Cruchet in Paris and Baron Constantin von Economo in Vienna, identified the condition's principal manifestation–its profound effect on sleep, that restorative period during which the body is at rest and the brain disconnects from the external environment. It was the clinical acumen of the aristocratic von Economo, a professor of psychiatry and neurology of Greek origin, whose masterful monographs on the subject gave us the best description of the disease and its underlying pathology of encephalitis, an inflammation of the brain.

Encephalitis lethargica (EL), epidemic encephalitis or von Economo disease starts with sore throat, nausea, headache, joint pain and fever–the general malaise associated with flu. From there the illness progresses to abnormal eye movements (which often cause double vision as the two eyes are no longer coordinated), drooping eyelids, an inability to open the eyes, and an irresistible need to sleep, day and night. This lethargy lasts for weeks, even in some cases for a year or more. Eleanore Carey, an eloquent young woman living in New York City, fell into a Sleeping Beauty–like trance in February 1923. Many years later she described these symptoms in a magazine article:

> After two months of illness I was in little pain, in fact, I was very comfortable, provided they did not prod me nor stand me on my head, turn me over in bed nor dash cold water on my face to waken me. It was so heavenly just to be allowed to sleep, but these people around me seemed determined to prevent my being comfortable! When the idea finally crept through my sleeping brain that I must waken, it seemed to be a physical impossibility. I wanted to be obliging, but I just could not. It seemed to me to be just as difficult to come to consciousness as it would have been had I been buried in a pit as deep as the center of the earth, where the circular walls about me were of shiny, polished marble. There were no crevices for my fingers on its sides nor any places to put my feet, but I must climb out of that pit with my bare hands! ... Perhaps it will give the reader a vague inkling of the dreadful lethargy which completely overpowers the victim of this disease and renders him impotent to make the effort to help himself.

These periods experienced by Carey and others came replete with vivid dreams, hallucinations and nightmares. When patients were woken up by calling their name or shaking them, they responded intelligently before quickly succumbing to somnolence again, often before they could fully provide answers to questions asked. The victims could not feed or clean themselves. In severe cases, they were stuporous or even comatose.

EL is a protean disease with some victims showing the opposite pattern, suffering from extreme insomnia or from inverted sleep–drowsy and lethargic during the day but delirious at night.

Mortality in this acute phase of the disease ran as high as 30 to 40 percent. Medical science was powerless to intervene and had to let the disease run its natural course. Hand in hand with EL's impact on sleep were its variegated effects on motor behavior, with patients displaying a menagerie of tics, motor and linguistic compulsions, tremors, rigidities and unconscious activities known as automatisms.

From Europe the disease spread to the Americas by way of New York and other port cities on the Atlantic coast. Estimates for the number of people worldwide who contracted EL ranged from about 52,000 reported cases upward to one million. The uncertainty derives from not knowing how many infected people experienced mild and short-lasting symptoms of EL that went unreported.

Why the epidemic strain of EL disappeared 10 years later as abruptly as it did remains unknown. Since then, only a handful of sporadic cases have been reported. Medical sleuthing has revealed previous instances of outbreaks of an EL-like disease, in particular in Tübingen, Germany, in 1712 (where it was dubbed *die Schlafkrankheit*) and in northern Italy in 1890 (*la nona*).

It is disconcerting that to this day, the culprit responsible for EL remains unknown and, presumably, at large. As epidemic encephalitis partially overlapped with the 1918–1920 epidemic of influenza–the H1N1 or "Spanish flu" that killed an estimated 50 million people–some consider the former a consequence of the latter. Yet no concrete causal evidence of any such link to influenza has surfaced. Nor did pathologists isolate from the brains of EL patients any bacteria or viruses that could be identified as the responsible agents.

A currently popular hypothesis suggests that the initial cold or flu virus triggered an inappropriate immune response. If true, EL would join an ever growing list of autoimmune afflictions. This hypothesis, however, fails to explain how the epidemic form of EL could burst onto the world stage, affect tens of thousands and then vanish. Most likely multiple factors were responsible, a set of antecedents that might in time converge again under the appropriate conditions to haunt humankind.

Unfortunately, worse was in store for some of the patients who survived the acute phase of EL. They subsequently developed Parkinson's disease to a varying degree, including an extreme form of akinesia, in which they remained immobile for decades, frozen statues locked away in nursing homes or wards for the terminally demented. These long-forgotten postencephalitic parkinsonian patients were the ones that neurologist Oliver Sacks awakened with L-DOPA drug therapy in the late 1960s, as famously described in his 1973 masterpiece *Awakenings*. The book became an eponymous movie with Robin Williams playing a character based on Sacks and an amazing, heart-wrenching performance by Robert De Niro portraying the patient Leonard L.

Sleep Is Actively Regulated

Historically the clinic has always served as the most fecund source of insight about the brain and the mind. EL remains a case in point. Most of von Economo's patients had eye-movement abnormalities and slept excessively. When awakened, they interacted relatively normally with their environment but soon returned to sleep. That is, EL has specific symptoms that reflect that the underlying disease does not destroy gray matter willy-nilly but selectively attacks particular sites for unclear reasons.

Postmortem examination of brains of patients invariably reveals discrete lesions in the midbrain, a structure that aptly describes its own whereabouts. It lies above the pons and just below the thalamus, the quail-egg-shaped structure that is the gateway to the neocortex, the rumpled outer layers responsible for perception, memory, thought and consciousness. The damage often includes the oculomotor nucleus in the midbrain, explaining the pronounced eye-movement dysfunctions characteristic of EL.

Damage also occurs in another deep-brain structure, the substantia nigra, part of the basal ganglia, an explanation for the patients' behavioral and motor pathologies. Von Economo proposed the existence of a structure in the midbrain that keeps the overlying thalamus and the neocortex in an activated, awakened state. This ascending arousal system sends fibers upward to the forebrain.

A smaller number of von Economo's EL patients showed the opposite reaction. They were able to sleep fitfully only for a few hours, beset by drowsiness most of the day but thwarted from sleep at night. These insomniacs had lesions in the gray matter of the anterior hypothalamus (literally, "below the thalamus"), extending into the basal ganglia. The physician inferred the existence of another center controlling the onset and depth of sleep. He argued that sleep is not just the passive response of a run-down body, tired after a day's work, ready to replenish itself. Rather it is an active state initiated and controlled by the tightly synchronized activity of specific organs in the central nervous system that are responsible for the daily choreography of our sleep-wake cycle.

Refined animal and clinical research over the past 80 years since von Economo published his work has validated his ideas and deepened our understanding of sleep and waking. We now know that there are two quite distinct forms of sleep, rapid eye movement (REM) and non-REM (or deep sleep). We know that circadian rhythms present in every cell in our body drive this system and are synchronized by the waxing and waning of daylight. The notion of a monolithic activating system has given way to the realization that sleep and arousal are controlled by activity within three dozen or more highly heterogeneous nuclei with idiosyncratic cell structures housed within the brain stem (that is, the medulla, the pons and the midbrain), as well as the hypothalamus and the basal forebrain.

It is a complex system that uses a variety of neurotransmitters: acetylcholine, noradrenaline, GABA, histamine, serotonin and orexin. Collectively they implement something akin to a bistable flip-flop, a type of electronic circuit in which two intrinsically excitatory networks mutually inhibit each other–when one is "on," the other one is "off," and vice versa. The system can be in one of two states—awake or asleep—with relatively abrupt transitions between the two.

When we lose the ability to sleep, to mercifully slip into its oblivion and forget our daily toils and troubles, we realize how utterly dependent we are on Morpheus' power. Indeed, forced sleep deprivation is rightfully considered a form of torture.

Let me end with Haruki Murakami, who perfectly captures the sensations of falling asleep in the final phrase of his 2013 novel *Colorless Tsukuru Tazaki and His Years of Pilgrimage*:

> The rear light of consciousness, like the last express train of the night, began to fade into the distance, gradually speeding up, growing smaller until it was, finally, sucked into the depth of the night, where it disappeared. All that remained was the sound of wind slipping through a stand of white birch trees.

Sleep well tonight.

Referenced

Hypothalamic Regulation of Sleep and Circadian Rhythms. Clifford B. Saper, Thomas E. Scammell and Jun Lu in *Nature*, Vol. 437, pages 1257–1263; October 27, 2005.

Encephalitis Lethargica: During and after the Epidemic. Edited by Joel A. Vilensky. Oxford University Press, 2010.

About the Author

Christof Koch is chief scientist of MindScope at the Allen Institute in Seattle and of the Tiny Blue Dot Foundation in Santa Monica, as well as author of The Feeling of Life Itself—Why Consciousness Is Widespread but Can't Be Computed. *He serves on* Scientific American'*s board of advisers.*

Section 2: The Secret to Better Sleep

Why Do Some People Need Less Sleep? It's in Their DNA

By Karen Weintraub

We all wish we could get by on less sleep, but one father and son actually can–without suffering any health consequences and while actually performing on memory tests as well as, or better than, most people.

To understand this rare ability, researchers at the University of California, San Francisco, first identified a genetic mutation–in both individuals–that they thought might deserve the credit. Then the scientists intentionally made the same small genetic spelling mistake in mice. The mice also needed less sleep, remembered better and suffered no other ill effects, according to a study published today in *Science Translational Medicine*.

Although a medication with the same benefits will not be available anytime soon–and might never materialize–the idea is incredibly appealing: take a pill that replicates whatever the father and son's body does and sleep less, with no negative repercussions.

"I find the concept of a gene product that might potentially provide protection against comorbid disorders of restricted sleep tantalizing," says Patrick Fuller, an associate professor of neurology at Harvard Medical School and Beth Israel Deaconess Medical Center in Boston, who was not involved with the work. "If true, this would indeed have 'potential therapeutic implications,' as well as provide another point of entry for exploring and answering the question 'Why do we sleep?' which remains [one] of the greatest mysteries in neuroscience."

But as Jamie Zeitzer, an associate professor in the department of psychiatry and behavioral sciences at Stanford University, notes, "There often are trade-offs." Zeitzer says he worries that even if a drug like this could be produced without causing significant side effects, it would still have social consequences. Some individuals might be forced or pressured to take medication so they could work more hours. Even if people will not need as much sleep, they will still need downtime, he insists.

The study's senior author, Ying-Hui Fu, a professor of neurology at U.C.S.F., says it is far too early for such fantasies. Instead she is interested in better understanding the mechanisms of healthy sleep to help prevent diseases ranging from cancer to Alzheimer's.

"These people sleep more efficiently," she says of the father-son pair. "Whatever function sleep is doing for us, it takes us eight [hours to feel rested], but it takes them six or four hours. If we can figure out why they are more efficient, we can use that knowledge to help everybody to be more efficient."

The subjects, who live on the East Coast, reached out to Fu's team after hearing about a previous publication of its work. She would not reveal any more information about them to protect their privacy, except that they are fully rested after four to six hours of sleep instead of the more typical seven to nine. Also, Fu says, the duo and others with similar mutations are more optimistic, more active and better at multitasking than the average person. "They like to keep busy. They don't sit around wasting time," she says.

If most people sleep less than their body needs, that deficit will affect memory and performance, in addition to measures of health, Fu notes. Many think they can get away with five hours of sleep on weeknights and compensate for the loss on weekends–but few actually can. "Your perception is skewed, so you don't really know your performance is not as good," she says. "That's why people think [adequate sleep] doesn't matter. But actually, it does. If you test them, it's obvious."

Joking about her own academic experience, Fu adds, "All those nights that I stayed up to study, it would have been better to go to sleep." That's not true of the father and son, who genuinely needed just 5.5 and 4.3 hours of sleep each night, respectively, the new paper showed.

Stanford's Zeitzer praises the study's design, saying, "Starting with humans and going to rodents and then back is great." Mice, he adds, are not ideal role models because they regulate sleep differently than humans. And many individuals believe they are short sleepers but, when put in a lab, turn out to slumber the typical seven to nine hours.

People are naturally short sleepers if they rest a relatively brief time even when given the chance to sleep in on weekends or vacations. "If you get extra sleep when you have the opportunity, it's generally a good sign that you need more sleep," Zeitzer says.

Jerome Siegel, a professor of psychiatry at the University of California, Los Angeles, Center for Sleep Research, says he is comfortable with Fu's group's main finding: that the *neuropeptide S receptor 1 (NPSR1)* gene is important in regulating sleep. But it is likely only one small piece in a very complex process, he adds. And he is not convinced by the connection between sleep and memory the group claims. Sleep may have many functions, but there is no indication, he says, that needing less of it somehow boosts memory or cognition. "We consolidate memory while we sleep and while we're awake, even when we're anesthetized," he says. "It's not something that just occurs during sleep."

The mechanism of action of the newly discovered mutation is not entirely clear. Fu and her team used a molecular probe to explore how the protein made by the father and son's mutant NPSR1 gene differs from that made by a normal gene. The mutation, they found, makes the receptor more sensitive and active. The specifics of that process, Fu says, still have to be worked out.

Fu and her collaborators previously discovered two other genes involved in sleep. They are continuing to explore the mechanisms behind these genes, she says, adding that the speed of their work would be faster if they had more financial support.

Fu says once she and her colleagues can find about 10 pieces of the genetic puzzle, "each piece can serve as a point to build upon. And hopefully, someday we can know the whole picture."

About the Author

Karen Weintraub is a staff writer at USA Today, *where she covers COVID, vaccine development and other health issues.*

Should We Make Daylight Saving Time Permanent? Let's Sleep on It

By Randyn Charles Bartholomew

The U.S. Senate has passed Florida Senator Marco Rubio's bill to make daylight saving time (DST) permanent across the country, which would end the practice of turning clocks forward in spring and back in autumn. For the Sunshine Protection Act to become law, the House of Representatives would need to approve it, and President Joe Biden would have to sign it, but the Senate passed the measure unanimously, signaling strong legislative support.

Scientists are less bullish on the idea. Our wakefulness is governed by a circadian rhythm inside us linked to the solar cycle. Although many sleep researchers approve of ending the clock changes, they prefer the use of standard time. The American Academy of Sleep Medicine issued a statement the same day Rubio's bill passed that said it "cautions that making daylight saving time permanent overlooks potential health risks that can be avoided by establishing permanent standard time instead." It added that "today's quick action by the Senate allowed for neither a robust discussion nor a debate. We call on the House to take more time to assess the potential ramifications of establishing permanent daylight saving time before making such an important decision that will affect all Americans." According to the academy's 2020 position paper in the *Journal of Clinical Sleep Medicine*, DST "is less well-aligned with intrinsic human circadian physiology, and it disrupts the natural seasonal adjustment of the human clock."

Less sunlight in the morning makes it harder for us humans to get started in the day, and more sunlight in the evening makes it harder to get to sleep. Darkness is a signal to the pineal gland in our brains that it's time to start producing more melatonin, which is our body's cue to lower internal temperature and start feeling sleepy. Early morning light is detected by the suprachiasmatic gland, which sits above the optic nerves, and its instructions cause our bodies to stop melatonin production so we can feel wakeful throughout the day.

The misalignment between the time on the clock during DST and our circadian rhythms means work and school occur at times that are not well suited to our biology, a phenomenon sleep researchers call social jet lag. "Studies show that social jet lag is associated with an increased risk of obesity, metabolic syndrome, cardiovascular disease, and depression," the position paper reported. Teenagers especially would suffer, it stated, because their circadian rhythms are delayed compared with other age groups, and shifting school and work obligations an hour earlier would run even more counter to their biology than it would for the rest of us.

In support of the Sunshine Protection Act, a press release on Rubio's Website cites studies from the *American Journal of Public Health* and the *Journal of Safety Research* showing that DST "reduces car crashes and car accidents involving pedestrians." But it doesn't mention that in the winter of 1973–74 when the U.S. stayed with DST, more children were killed by cars in the early mornings as they were heading to school in the dark.

The press release quotes Florida Senator Rick Scott, a former governor, saying, "As our state works to fully reopen and bring visitors back safely, this legislation will give families more time to enjoy all that Florida has to offer." Indeed, the golfing and hospitality industries favor DST. So does the Association for Convenience and Fuel Retailing, which advocates for roadside gasoline stations and convenience stores. Studies show that gasoline sales are higher when it's light later, and people play more golf, so these industries are willing to back the bill, even if many scientists believe it compromises Americans' sleep health.

Bill co-sponsor Ed Markey, senator from Massachusetts, claims the act would "add an extra hour of sunshine for the full 365 days a year." Sadly, the actions of the sun are well outside the scope of congressional authority. Every day our terrestrial sphere revolves into and out of the sun's light, irrespective of what we on its surface codify into laws and regardless of what numbers we display on our clocks.

DST shifts an hour of sunlight from the morning to the evening, which during the winter solstice in New York City would move sunrise from 7:10 a.m. to 8:10 a.m. This change would give most people an extra hour of evening light at the cost of an extra hour of morning darkness. A better name than daylight saving time might be daylight shifting time. DST is merely putting different numerical labels on the solar events of each day. To think we can alter our circadian rhythms that way would be like trying to lose weight by converting to the metric system, or cool off on a summer day by switching thermometers to Celsius.

The very legislators who support the permanent change often seem confused on this point. CBS News quoted Senator Roy Blunt of Missouri as stating, "I just think the extra hour at the end of the day consistently is better than having it dark when kids go to school and dark when kids get home." Again, we are not actually creating more daylight here. That would be some mighty legislation indeed.

The U.S. has been in this position before. In December 1973 Congress implemented permanent daylight saving time in response to an Organization of Arab Petroleum Exporting Countries oil embargo, with the hope that it would save energy. The idea was well-loved at the outset, earning a 79 percent approval rating in polls. The rating dropped precipitously to 42 percent once people experienced their first winter on the new time; for many of them, DST during winter involved rising before the sun did.

Ironically, it is unclear if permanent DST even saved on energy costs. Although studies make positive and negative cases on the issue, a 2011 paper from MIT Press's *Review of Economics and Statistics* found that when DST was implemented in Indiana, "contrary to the policy's intent, DST increases electricity demand." The paper cited a trade-off between a decrease in power needed for lighting and an increase needed for heating and cooling.

Some human religions, past and present, typically placed a sun deity at the top of their mythological hierarchy. Although modern folks, especially those of us who live in cities, no longer have the same intimate relationship with the sun as our ancestors did, our biology is still tied to its influence.

Pulling ourselves out of bed in a predawn gloom is more than unpleasant; it can have far-reaching health effects. A 2019 joint statement by the European Biological Rhythms Society, the European Sleep Research Society and the Society for Research on Biological Rhythms said that using standard time "improves our sleep and will be healthier for our heart and our weight. The incidence of cancer will decrease, in addition to reduced alcohol and tobacco consumption. People will be psychologically healthier, and performance at school and work will improve."

About the Author

Randyn Charles Bartholomew is a freelance writer, political news junkie and math enthusiast living in Astoria, Queens. Follow him on Twitter @rcbarthol.

The Secret to a Better Night's Sleep: A Sense of Purpose?

By Daisy Grewal

Despite its importance for health and well-being, many American adults find it difficult to consistently get enough sleep. Approximately 50 million to 70 million Americans suffer from a sleep disorder, according to the Centers for Disease Control and Prevention. Sleep disturbances are particularly common in older adults and involve a variety of problems including difficulties falling or staying asleep, interrupted breathing and restless leg syndrome. A person's racial background can influence his or her likelihood of developing a sleep disorder, with a greater number of Black Americans reporting sleep disturbances compared to white Americans.

Beyond its effects on health, not getting enough sleep can lead to car accidents, medical errors or other mistakes on the job. To encourage better sleep, the medical community encourages adults to engage in good "sleep hygiene" such as limiting or avoiding caffeine and nicotine, avoiding naps during the day, turning off electronics an hour before bed, exercising and practicing relaxation before bedtime. It is also well-known that mental health is closely linked to sleep; insomnia is more common in people suffering from depression or anxiety.

A recent study now raises the possibility that sleep could be affected by the degree to which someone feels like his or her life is purposeful or meaningful. Arlener Turner, Christine Smith and Jason Ong of the Northwestern University School of Medicine found that people who reported having a greater sense of purpose in life also reported getting better sleep–even when taking into consideration age, gender, race and level of education.

To establish this link, the researchers recruited a sample of 825 older Americans to participate in a study where they reported on their sense of purpose in life along with the quality of their sleep. The majority of these participants were female (77 percent), and slightly more than half were African-American (54 percent). The participants were, on average, 79 years old. A sense of purpose in life was measured using a survey where participants rated how much they agreed with each of 10 statements, such as "Some people wander aimlessly through life, but I am not one of them." The results showed that participants who reported having a greater sense of purpose in life also reported higher quality sleep on a regular basis, as well as fewer symptoms of sleep disorders. Importantly, the researchers found that their findings held true for both the white Americans and Black Americans who participated in the study.

It is important to emphasize that this study only looked at the *association* between a sense of purpose and better sleep–the findings cannot say for sure that having a greater sense of purpose causes one to sleep better. An alternative interpretation for the findings is that people who have a greater sense of purpose also tend to have better physical and mental health, which in turn explains their higher quality sleep. Another important limitation of the study is that the findings rely entirely on people's self-reported sleep symptoms. The researchers did not bring participants into a lab and actually monitor the quality of their sleep. Therefore, it is possible that people with a higher sense of purpose simply remember getting better sleep compared to people who do not report experiencing a sense of purpose in life.

Despite these limitations, this study is the first to suggest any kind of strong link between purpose in life and sleep. Given how common sleep problems are, anything that may suggest new avenues for treatment is important to explore. Perhaps developing a sense of purpose in life could be as effective at improving sleep as following healthy habits, such as limiting coffee. In addition to promoting good sleep hygiene, doctors may end up recommending mindfulness practices or exploring one's values as ways of helping older adults sleep better. Given how elusive a good night's sleep has become for many, it's well worth exploring. The impact of poor sleep goes far beyond our own personal health, as the side effects have the potential to wreak havoc on other people's lives as well.

Developing a sense of purpose in life may simultaneously convey other benefits in addition to better sleep. Research has linked experiencing purpose in life to a variety of other positive outcomes including better brain functioning, reduced risk of heart attack and even a higher income. People with a greater sense of purpose in their life would surely be better off while also serving as a positive example in the lives of those they know.

About the Author

Daisy Grewal holds a B.A. in psychology from the University of California, Los Angeles, and a Ph.D. in social psychology from Yale University. She is a research scientist on the people analytics team at Intuit in Mountain View, Calif.

Good Snooze for Insomniacs–Therapy Is as Effective as Sleep Drugs

By Jessica Schmerler

Imagine the following scenario: You have a big presentation at work tomorrow and know you need a good night's sleep tonight so you don't look tired or forget your lines. You get to bed at a reasonable hour but can't turn off your anxiety and begin mentally berating yourself about your inability to fall asleep, which only prolongs your wakefulness. Eventually, you get frustrated and hit the medicine cabinet so you can finally get some much-needed shuteye.

Sound familiar?

One in three adults report problems with insomnia, and when you're feeling stressed or anxious, it is easy to pop some sleep aids for a quick fix every so often. Usually this helps, but for the 5 to 15 percent of adults suffering from chronic insomnia, relying on prescription pharmaceuticals such as benzodiazepines and related "Z drugs" (such as Ambien or Lunesta) is not an attractive option. These compounds can present a risk of chemical or psychological dependence, and patients often develop a tolerance to their effects over time. As part of efforts to provide patients with more options, some doctors lately have proposed psychological interventions such as cognitive behavioral therapy for insomnia (CBT-i).

CBT is a kind of talk therapy used to treat a variety of mental illnesses. At its foundation is a model known as the "vicious circle," also known as spiraling. The vicious circle explains how negative thoughts lead to feelings and then to self-destructive behaviors, and so on. CBT is a technique that intervenes when these negative thoughts start, with the goal of influencing the behavior that follows.

For insomnia, treatment is administered in five phases: Therapists prompt their patients to identify negative beliefs about sleep that might contribute to insomnia, such as an irrational fear of not getting enough sleep, and then explain why these ideas are unhelpful or suggest alternative, more positive thoughts. In the second part of the treatment, stimulus control, therapists try to maximize the association of the bed with sleep by instructing patients on behavioral changes such as avoiding stimulating activities in the bedroom. Third, they advise patients to go to bed only when sleepy, with the idea of minimizing the amount of time there spent lying awake. Fourth, they make recommendations on how to keep good "sleep hygiene," which refers to environments and habits that promote sound sleep, such as avoiding daytime naps and limiting caffeine intake before bed. Finally, therapists suggest patients try relaxation techniques such as meditation and mindfulness just before bed to try to quiet their racing minds.

Physicians at the Melbourne Sleep Disorders Center (MSDC) in Australia recently undertook a systematic review and meta-analysis of studies pertaining to the efficacy of CBT-i in the treatment of chronic insomnia. James Trauer, a sleep physician at MSDC and lead author of the paper, says he undertook the review because he was surprised by how few of his patients have tried or even heard of CBT-i, "despite it being the most effective technique, as it gets at the core of the problem, which is frustration with time awake in bed." Based on the results of the studies they analyzed, the authors concluded that CBT-i is effective and a good alternative to pharmacological intervention for the treatment of chronic insomnia. Their findings were published in *Annals of Internal Medicine* on June 8.

Out of 292 citations and 91 full-text articles, the authors selected 20 trials that met their selective criteria–ruling out studies that, for example, incorporated fewer than three of the five components of CBT-i or only included patients that had other medical conditions as well as insomnia. These studies measured four determinants of sleep quality: time taken to fall asleep, time spent awake throughout the night after initially falling asleep, total sleep time and sleep efficiency (the total sleep time divided by the total recorded time). The authors reported the average of the results of all 20 studies, which showed that after participating in a course of CBT-i, patients fell asleep 19 minutes earlier, spent 26 fewer minutes awake in the middle of the night, got 7.6 more minutes of sleep overall, and improved their sleep efficiency by almost 10 percent.

Although the change in total sleep time was not statistically significant, the remaining three measures showed marked improvement. Based on these statistics, the authors reported that CBT-i resulted in similar levels of improvement to those measured in trials of benzodiazepines for treating insomnia but without the risk of side effects associated with such pharmaceuticals. The authors did not include comparisons with trials of Z drugs, despite their rise in popularity in the past several years. Because of the recentness of their development, data on their long-term effects and safety is limited.

Of greatest significance, however, was that the effects of CBT-i were found to be more sustainable in the long run than those of pharmacological interventions. Charles Morin, a cognitive neuroscientist at Laval University in Quebec who was not involved in the study, points out that CBT-i provides a more lasting solution than pills alone because it addresses the underlying psychological and behavioral factors that perpetuate insomnia over time. This means the CBT-i option, compared with a lifetime of taking pharmaceuticals, "is likely to be a cost-beneficial investment in the long run," Morin says.

Although the findings have implications for managing insomnia's symptoms, there are caveats. The studies included in the review only measured quality of sleep and did not address the impact of CBT-i on quality of life and other factors. Insomnia has consequences beyond the seven or eight hours spent in bed, including health problems, psychological distress, and economic burden due to lost productivity.

In addition, pursuing CBT-i requires more time and motivation than taking a pill before bedtime. Sleep medications are attractive because they allow insomniacs to take one pill before bed that helps them fall asleep and stay asleep without any effort on their part. With CBT-i, one of the recommendations of stimulus control, for example, is leaving the bed if unable to fall asleep after 15 minutes and trying again later. Studies establishing that CBT-i directly improves quality of life would be required to show CBT-i is truly a better alternative to all pharmacological intervention.

If CBT-i proves to be the best option for troubled sleepers, researchers will also need to find ways to spread the word about this treatment. In contrast to the marketing campaigns common among pharmaceutical companies, behavioral treatments often go unnoticed. And, even with awareness of CBT-i, patients may not have the time, money or inclination to work with a psychotherapist on a regular basis.

One solution is an online app called Sleepio, developed by University of Oxford neuroscientist Colin Espie, that allows people to input their sleep patterns and negative thoughts about sleep, then generates a personalized CBT-i plan. The program "rings a bell with people because it is engaging, imaginative *and* evidence-based," Espie says.

The MSDC meta-analysis comes at a time when society has become highly aware of the heavy physical and emotional burden of insomnia. Chronic insomnia puts people at risk for countless health conditions, including hypertension, diabetes, and depression, as well as increases the likelihood of substance abuse and vehicular accidents. Indeed, the studies provide evidence of CBT-i's efficacy but they also provide sufferers with a choice as to how they want to manage their care.

"The most important message for patients," Trauer says, "is that there is an effective treatment out there that is safe, doesn't rely on popping pills, and gets to the core of the problem."

Putting Insomnia on Ice

By Joe Kloc

The pain and frustration of chronic insomnia affects one in 10 American adults, most of whom find no relief from current therapies. Now a new study finds that simply cooling the brain area just behind the forehead can help.

In a study presented this summer at the American Academy of Sleep Medicine's SLEEP 2011 conference, researchers fit 12 insomniacs with caps that use circulating water to cool the prefrontal cortex. The cap helps the insomniacs fall asleep about as fast—and stay asleep about as long—as adults without insomnia.

"When you get into the neurobiology, insomnia is a disorder of hyperarousal," says Eric A. Nofzinger, a psychiatrist at the University of Pittsburgh School of Medicine who worked on the study. In adults with normal sleeping patterns, the metabolism of the prefrontal cortex decreases as they fall asleep. In insomniacs, however, it increases—corresponding with the incessant worrying or brain chatter that many insomniacs report experiencing. Using the cap to perform a cooling process on the brain called cerebral hypothermia, the researchers were able to reduce the brain's activity and lull the subject to sleep.

The finding is significant because current treatments such as hypnosis and sleeping pills help only about one in four insomniacs. The cooling cap, which had a 75 percent success rate, may soon offer patients a safe, comfortable, nonpharmaceutical way to enjoy a good night's sleep. Participants reported that wearing the cap was a "soothing, massagelike experience," Nofzinger says. "Imagine your grandmother putting a cold washcloth on your forehead." He hopes that the cap may also prove useful to patients with anxiety and mood disorders, which also involve the prefrontal cortex.

How Do CPAP Machines Treat Sleep Apnea?

By Meghan Bartels

President Joe Biden is using a machine to keep his airways clear while sleeping because he has a condition known as sleep apnea, White House officials say.

Sleep apnea likely affects about 30 million people in the U.S., according to the American Medical Association. Some people with sleep apnea use what's called a continuous positive airway pressure (CPAP) machine, which includes a mask that a user straps onto their face at night to breathe better while they sleep.

Officials revealed Biden has been using a CPAP machine after reporters recognized indentations on his face from the strap holding the device in place. Documents released as early as 2008 show Biden has been known to experience sleep apnea, which can become more common as people age.

Scientific American spoke with Christine Won, medical director of the Yale Centers for Sleep Medicine, about sleep apnea, its causes and treatment with CPAP machines.

[An edited transcript of the interview follows.]

Bartels: What is sleep apnea?

Won: Obstructive sleep apnea occurs because there's a problem with the structure or the function of the upper airway. When we fall asleep, the upper airway loses its muscle tone, and if your airway is narrow, this might lead to obstruction. This can cause people to snore, or it can also completely obstruct the airway and cause people to not be able to ventilate well. This can lead to low oxygen levels; it can lead to disturbances in sleep. People can sometimes awake abruptly as a result, and sleep is fragmented.

Bartels: Are there different kinds of sleep apnea?

Won: Yes, [in addition to obstructive sleep apnea, there is a type called] central sleep apnea, which is far less common. It generally affects older men and people with heart disease, although it can be found sometimes in healthy individuals. This type of sleep apnea results from the brain not communicating during sleep to the respiratory muscles to take a breath, and so there are pauses in breathing.

Bartels: What are the symptoms of sleep apnea?

Won: Both types of sleep apnea can disrupt sleep because the person may awake abruptly because they're not getting enough oxygen. Snoring comes more with the obstructive sleep apnea. And then both types, because they can disturb sleep, can lead to daytime symptoms such as sleepiness or brain fog, these types of things.

Bartels: How common is sleep apnea?

Won: Sleep apnea is very common in the adult population. Prevalence increases with weight gain–so it's much more prevalent in individuals with obesity. It's more common as we age. It's also twice as common in men as in women, for unclear reasons.

There are a lot of people who have sleep apnea that are undiagnosed. I think current data suggest only about 20 percent of people with sleep apnea are actually diagnosed.

Bartels: Why do you think so many people go undiagnosed?

Won: Probably a little bit of education–not knowing what the symptoms of sleep apnea are, not knowing that waking up a lot at night, unrefreshing sleep, these types of things can be a sign of sleep apnea and not just, "normal." There might be a little bit of resistance to undergo a sleep study, and maybe there might be a little bit of resistance in accepting CPAP as a form of therapy for this.

Bartels: What causes sleep apnea?

Won: The structure problem is that the airway can be narrowed for various reasons. There could be excessive fat in the neck tissues.

The soft palate could be redundant and low-hanging and basically blocking the airway. The tongue base could be very big and bulky, again, reducing the airway size. Some people can have very large tonsils and adenoids [lymphatic tissue near the tonsils] that can obstruct the airway. Also, the jaw anatomy can affect the airway size–people who have really small jaws or recessed jaws, their airway can be narrowed.

And some people can have very wide, open airways while they're awake, and they don't have any of those anatomical issues that I just described. But when they fall asleep, airway muscle tone decreases markedly and can lead to airway collapse. It's not clear why. But the upper airway is basically a large muscular tube: it's not supported by bony structures, so muscle tone is important to keep the integrity of the airway. That may be why aging is related to increased risk of sleep apnea, because with aging, all our muscles get a little bit more relaxed.

Bartels: What are common treatments for sleep apnea?

Won: CPAP is the most common treatment for sleep apnea because it's highly efficacious–so in other words, if you wear it, almost always it's going to eliminate obstructive sleep apnea. The problem is that a lot of people have difficulty tolerating CPAP, so it's not an effective treatment for everyone.

Other treatments include oral appliances that you get fitted for by your dentist. You fit the appliance into your mouth at night, and it pulls your lower jaw forward so the airway doesn't collapse while you're sleeping. That's a reasonable treatment option for people who are young and thin and generally have mild sleep apnea.

There are also a lot of surgeries, such as a maxillomandibular advancement surgery, which is a pretty big surgery where they cut your jaw and advance it–up to a centimeter sometimes–to open up the airway in the back. Tonsillectomy and adenoidectomy are actually mainstay treatments for children with sleep apnea–generally not for adults.

A newer form of therapy is upper airway stimulation. It's like a pacemaker for the tongue, essentially. But it's not a therapy for everyone–there are specific criteria and a specific population who would benefit from that.

And not all sleep apnea needs to be treated. We know that sleep apnea is associated with cardiovascular disease and risks. What we don't know is how much CPAP therapy benefits [people or who will benefit from it]. So even if you're diagnosed with sleep apnea, it doesn't automatically mean that you go to CPAP therapy. You'd have a discussion with your doctor about [questions such as] "How symptomatic are you? What kind of cardiovascular risks do you have? How severe is your sleep apnea?" Based on these things, your doctor will decide if you would benefit from treatment or not.

Bartels: How do CPAP machines work?

Won: They're basically reverse vacuum cleaners. They have a motor that generates positive pressure that's transmitted via a tube and a mask into the pharyngeal space. And it acts like a pneumatic splint, and it just holds open the airway, so it doesn't collapse.

Bartels: How easy is it to get a CPAP machine?

Won: They're very accessible: if you're diagnosed with sleep apnea, and you meet specific criteria from an insurance perspective, then it's easy to prescribe and obtain one.

It is somewhat difficult for patients to keep their CPAP or get renewed supplies, because for insurance, you sometimes have to meet these metrics of using it at least four hours a night for 70 percent of nights for 30 days consecutively. It's kind of tough because a lot of people don't take to CPAP right away. Some people do–like first night, they love it; it's the best sleep of their life. But other people, it takes some time to get used to it, because this machine on your face is blowing pressure. It's a foreign sensation.

Bartels: What should people do if they think they might have sleep apnea?

Won: They should definitely discuss it with their doctor, who can further screen and evaluate their symptoms and decide if they are at risk for it and then refer them for a sleep study.

About the Author

Meghan Bartels is a science journalist and news reporter for Scientific American *who is based in New York City.*

What to Do When You Can't Fall Asleep May Surprise You

By Meghan Bartels

We've all been there: lying in bed wide awake and desperately wondering how to get to dreamland. In fact, scientists say it's pretty normal to have a little trouble falling asleep or staying asleep from time to time.

"There's this expectation that we should just go to sleep and stay sleeping for seven to eight hours," says Roxanne Prichard, a neuroscientist at the University of St. Thomas, Minnesota. "That's just not biologically supported with how humans sleep."

While having trouble drifting off to sleep isn't unusual, it still can be frustrating. Fortunately sleep experts are gaining an increasingly strong understanding of what's happening in the brain during the process–and they say you can use that knowledge to increase your chances of catching some z's, even when sleep feels elusive.

"It's very uncommon for people to be able to just transition from being awake and active to falling asleep right away," says Kim Hutchison, a sleep medicine specialist at Oregon Health & Science University.

Falling asleep is a big shift for your brain. When conditions are good, hitting the sack allows your brain activity to slow down and become more orderly, and your brain waves synchronize, Prichard says. This switch is governed in part by environmental cues, such as light or temperature. A hot summer night or a bright streetlamp seeping through the window can interfere. The process of nodding off can also be affected by your emotions.

"You need to feel both physically and psychologically safe to sleep," Prichard says. "If there's something that you are really worried about, if you are sleeping next to someone you don't trust, if you're worried that the newborn that you're caring for might stop breathing, it's going to be hard to fall asleep."

That's why anxiety and stress are key culprits when people are unable to fall and stay asleep—and why relaxation is a crucial tool for easing into slumber. Hutchison and Prichard both say that the key, whether you're struggling to sleep just as you're heading to bed or after waking up in the middle of the night, is to limit the amount of time you spend lying awake fretting about not being asleep.

"If you're unable to fall asleep in what seems like or feels like 20 minutes or so, or you feel your body getting more amped up because you're getting anxious that you're not falling asleep, then I would recommend getting out of bed and sitting somewhere quietly with dim light and just relaxing, doing something boring," Hutchison says.

You could use the time to read, listen to calming music, drink some chamomile tea or do breathing exercises—anything that slows and comforts your body and mind, Hutchison and Prichard say. Avoid snacking, exercise and screens.

It's also important to resist the urge to fixate on worries, concerns or challenges, both experts say, especially in the middle of the night. Not only will those thoughts keep you awake, you also probably won't make much progress on what's keeping you up because your prefrontal cortex, a part of the brain that excels at planning and analysis, doesn't get as much blood at night.

An emerging hypothesis suggests the brain isn't well suited for cognitive processing in the wee hours, Prichard says. "The mind, after midnight..., is less equipped to problem-solve and more prone to find problems," she explains. "It's really easy to wake up in the middle of the night and freak yourself out about something that you could probably problem-solve more effectively later."

Sleep aids and supplements, such as melatonin, are commonly used to fall asleep, but Prichard likes to advise people to use relaxation practices instead. "I want people to learn sleep skills, not pills," she says.

If basic relaxation strategies fail to calm your brain, both Prichard and Hutchison recommend cognitive-behavioral therapy for insomnia.

Even if your night's rest is incomplete, avoid the urge to make up for it by sleeping in or napping the next day, Hutchison says, because that can backfire. To fall asleep more quickly the next night and get back on track, it's important for your brain to crave sleep. "If you nap during the day, especially longer naps, then your brain gets little snippets of sleep and will be less likely to fall asleep quickly at bedtime," she says.

Although Hutchinson and Prichard emphasize that occasional sleep struggles are normal, they recommend seeing a doctor if the situation continues or if it interferes with your ability to function during the day–which could indicate something physiological is at play. "In general, if you're having problems falling asleep for three nights or more per week, then we would consider that problematic, but it also needs to interfere with your daytime functioning," Hutchison says.

For most sleep seekers, if something is keeping you up at night, it's no help trying to force yourself to sleep. The best trick may be to simply preoccupy your mind until slumber comes naturally.

About the Author

Meghan Bartels is a science journalist and news reporter for Scientific American *who is based in New York City.*

Section 3: Poor Sleep and Health

The Sleep-Dementia Connection

By Claudia Wallis

Among the many things that can shatter when Alzheimer's disease tightens its grip is the steady rhythm of the body's sleep-wake cycle. The problem is so common that one New York City nursing facility—the Hebrew Home at Riverdale—ran an all-night program for many years that took in afflicted community members for a dusk-to-dawn schedule of games, snacks, arts and crafts, and other activities so that their exhausted families could get some shut-eye.

Troubled sleep often begins long before dementia becomes apparent. In recent years research has been heating up on two key questions: Could disrupted sleep be a reliable early warning sign that the brain changes of Alzheimer's have begun? And even more exciting, though still speculative: Could the onset of the disease or its progression be slowed by treating sleep-related issues?

The brain pathology of Alzheimer's gets underway roughly 20 years before symptoms such as memory lapses and confusion become obvious. Scientists believe the fateful sequence goes something like this: beta-amyloid, a nerve cell waste product, starts to accumulate in the spaces around brain cells, eventually forming the telltale plaques of Alzheimer's. This is followed by a toxic buildup of tangles of tau protein inside nerve cells, first in the medial temporal lobe and then spreading to other regions. These changes lead to the death of neurons, loss of synapses and general atrophy seen in Alzheimer's-addled brains and the observable deterioration of cognition and behavior.

Sleep, as it turns out, impacts both beta-amyloid and tau. Studies in humans and mice indicate that levels of both proteins fall during sleep. People who sleep poorly have higher levels of beta-amyloid and tau in their cerebrospinal fluid—even after a single bad night. What is perhaps more significant is what happens over the long term. PET scans show older adults with chronic sleep problems have more beta-amyloid deposited in their brain. Research published earlier this year in *Science* revealed that in a mouse model of Alzheimer's, lack of sleep promotes the spread of abnormal tau across certain brain regions. "It suggests that if there's a sleep disturbance night after night for some prolonged period, it could expose an individual to higher concentrations of these proteins and increase the risk of Alzheimer's," says Brendan Lucey, one of the *Science* paper's authors and an assistant professor of neurology at Washington University in St. Louis.

A 2018 study provides evidence for this cumulative damage scenario. Using data on 124 older adults participating in a long-term National Institute on Aging (NIA) study, sleep researcher Adam Spira and his colleagues at Johns Hopkins University and the NIA found that people who complained of "excessive daytime sleepiness" at an average age of 60 were 2.75 times more likely to have beta-amyloid plaques in their brain some 16 years later.

In another new study, published earlier this year in *Science Translational Medicine*, Lucey and his colleagues explored which part of sleep may be most relevant to Alzheimer's pathology. They found that having less of a very deep phase called slow-wave, non-rapid eye movement sleep, is associated with more accretion of tau and, to some degree, beta-amyloid. This part of sleep also happens to be important to memory consolidation.

No one knows for sure what comes first: Do excess beta-amyloid and tau impair sleep, or does impaired sleep lead to a buildup of these proteins? The leading hypothesis is that it goes both ways in a kind of vicious cycle. For example, Lucey proposes, let's say someone "develops sleep apnea, sleeps poorly, increasing Alzheimer's pathology, which then worsens their sleep further and accelerates their pathology."

Could disrupting this cycle help stave off dementia? It's too early to say, but as Spira notes, "there's a growing interest in coming up with ways to attack poor sleep as a possible way to prevent Alzheimer's." A small study published earlier this year by a team at the University of California, San Francisco, offers a glimmer of hope. It looked at 50 older adults with Alzheimer's, mild cognitive impairment or normal cognition and found that the 25 who used the drug trazodone as a sleep aid had a significantly slower cognitive decline than the 25 who did not use the it. Trazodone, intriguingly, is known to increase slow-wave sleep. Alas, the road to an Alzheimer's treatment is littered with disappointment. Working the sleep angle may prove to be yet another pipe dream, but it's hard to imagine it could do much harm.

About the Author

Claudia Wallis is an award-winning science journalist whose work has appeared in the New York Times, Time, Fortune *and the* New Republic. *She was science editor at* Time *and managing editor of* Scientific American Mind.

Weekday–Weekend Sleep Imbalance Bad for Blood Sugar Regulation

By Karen Hopkin

Few things are more satisfying than sleeping late on weekends. But though the extra z's may improve your mood, they do not appear to improve your health. Because a new study shows that so-called "recovery sleep" cannot reset the body's metabolic clock ... and may actually lead to some serious health issues.

"Sleep loss can impact a range of physiological systems. It can increase our risk for cardiovascular disease, it can cause weight gain, it can decrease insulin sensitivity, so it can increase our risk of diabetes," said Christopher Depner, an assistant professor in the department of integrative physiology at the University of Colorado, Boulder. "And this can happen, depending on what you're looking at, this can happen as quick as one or two nights of not getting enough sleep."

A lot of us shut off the alarm on Saturday and Sunday. But we go right back to burning the candle at both ends once the workweek begins.

"So we were really interested in how this sort of cycle of insufficient sleep, weekend recovery sleep, insufficient sleep, can impact your risk of obesity and metabolic disease."

Depner and his colleagues invited volunteers to a nine-day snooze-a-thon. One group was allowed to get a full night's sleep. The next was kept to just five hours each night. And the third group went back and forth, restricted to five hours of shut-eye during the workweek, allowed as much sleep as they wanted over the weekend, and then back to five hours for the last couple days.

The results?

"Well, the key findings from this study show that when we maintain insufficient short sleep schedules during a typical work or school week, we find that this leads people to eat more than they need and this leads to weight gain," said Ken Wright, professor of integrative physiology at U.C. Boulder. He's the senior author of the study, which appears in the journal *Current Biology*.

So it seems a loss of sleep leads to a spike in snacking. But even more surprising, sleeping in on the weekend doesn't help—and even makes things worse.

"We found that after the weekend, when they went back to getting insufficient sleep during the work or school week, we found that their liver and their muscle insulin sensitivity or blood sugar regulation was reduced. And this is not something we had found in people who maintained chronic insufficient sleep schedules. So it's possible that, yes, this is a worsening of the body's ability to regulate blood sugar for those specific tissues after the weekend."

So make a date with a pillow. And trade the sweets for sweet dreams.

[The above text is a transcript of this podcast.]

About the Author

Karen Hopkin is a freelance science writer in Somerville, Mass. She holds a doctorate in biochemistry and is a contributor to Scientific American's *60-Second Science podcasts.*

Sleep's Role in Obesity, Schizophrenia, Diabetes...Everything

By Gary Stix

Is sleep good for everything? Scientists hate giving unqualified answers. But the more sleep researchers look, the more the answer seems to be tending toward a resounding affirmative.

The slumbering brain plays an essential role in learning and memory, one of the findings that sleep researchers have reinforced repeatedly in recent years. But that's not all. There's a growing recognition that sleep appears to be involved in regulating basic metabolic processes and even in mental health. Robert Stickgold, a leading sleep researcher based at Harvard Medical School, gives a précis here of the current state of *sommeil* as it relates to memory, schizophrenia, depression, diabetes—and he even explains what naps are good for.

Stix: How far have we come in understanding sleep?

Stickgold: Although we understood the function of every other basic drive 2,000 years ago, we are still struggling to figure out what the biological functions of sleep are. One of the clearest messages now is that for every two hours humans spend awake during the day, the brain needs an hour offline to process the information it takes in and figure out what to save and what to dump and how to file and what it all means.

Stix: So what is sleep for?

Stickgold: Memories are processed during sleep. But sleep doesn't have just one function. It's a little bit like listening to tongue researchers arguing about whether the function of the tongue has to do with taste or speech. And you want to say: 'Guys, c'mon, it's both.' There's very good evidence now that sleep, besides helping memory, has a role in immune and endocrine functions. There's a lot of talk about to what extent the obesity epidemic is actually a consequence of too little sleep.

Stix: How did sleep come to play so many diverse roles?

Stickgold: It's a little bit of a tricky game because it looks like sleep has been identified evolutionarily as a good time to get stuff done. If you work in a big office, the cleaning staff comes in at night. It's not because that's when they like to or that's when they're most efficient but that's when the people are out of the office and the office is shut down and it's easiest to get housekeeping chores done. I imagine that some functions over evolutionary time have glommed onto sleep because that's when the system is most shut down. Something like two-thirds of your growth hormone is secreted during deep, slow-wave sleep and the presumption is not so much that that this is a critical function of sleep, but that this is the best time to pause and spend energy on growth.

Stix: What is sleep's relationship to memory?

Stickgold: Sleep plays a clear and often critical role in the stabilization of memories and in their strengthening and retention, allowing some memories to be retained and others to be let go. It's also critical in the extraction of gist from large collections of information and the discovery of rules that control that information so if you give various stimuli, sleep can help you determine what the rules are as to what the stimuli produce and can even help with the development of insight in patterns and rules that you didn't even know were there.

Stix: What is the importance of dreams?

Stickgold: Much less, much less. A decade back we had a paper in *The Journal of Cognitive Neuroscience* in which we showed that episodic memories of events in our lives aren't replayed in dreams. So the old concept that you dream *about* what happens, you don't dream what *actually* happened, is actually well validated. The dreaming brain seems to have as its goal to construct hallucinatory scenarios that bear resemblance to recent events but are different from them.

Erin Wamsley and I have shown with a computerized maze learning task that if you take a nap after learning you improve on that task more than if you stay awake. Also, if you collect dream reports during that nap, you find it's the people who report that they've been dreaming about the maze task who show the preponderance of improvement during re-testing. Interestingly, if you look at the dream content, it's very clear that the dream content *per se* is not helpful. It has a sort of wacky bizarre relationship to the task, but not in a way that would help you perform better, which has led us to describe it as a sort of biomarker of a memory process.

What we think we're seeing is that at the same time that the hippocampal region of the brain, which we know encodes spatial maps, is replaying and strengthening its memory for the layout of the maze, other regions of the brain involved in dreaming are constructing these 'what if' situations, imaginary future scenarios, sort of getting at the question of what is the usefulness of this information.

One subject reported dreaming of the maze and then remembering being in a bat cave a few years ago, and you can imagine his brain saying: 'OK, how do these two fit together? Is there something about this bat cave that might help me do this task better? Or is there something about this maze task that I should not just file under experiments that I participated in, but file in a way that if I go exploring in a cave again then maybe something I learned about exploring a maze from this game I just played might be useful?' And I think we're getting a sense that that's what dreaming's about–how might this information be useful to me in the future? It's exploring associative networks and trying to find associations that look promising.

Stix: What are implications of sleep for psychiatric disorders?

Stickgold: If you take an adult who has both sleep apnea and depression, you'll find that they are very tightly linked. If you have depression, there's a fourfold increase in your likelihood of apnea and if you have apnea, there's a fivefold increase risk of depression. If you take someone with both depression and apnea, and treat the apnea with CPAP [continuous positive airway pressure], you can get their depression scores to drop below clinical levels.

If you take kids comorbid for sleep apnea and ADHD–in the case of children the apnea is usually caused by enlarged adenoids and tonsils–if you remove the tonsils and adenoids you'll get a larger reduction in the ADHD symptoms than if put them on Ritalin.

If you take people with bipolar disorder and sleep deprive them, you'll flip them into the manic state.

If you look at depressed people, REM sleep comes much earlier in the night. When treatments for depression fail to reverse this effect, the likelihood of recurrence of the depression is much higher. And depriving depressed patients selectively of REM sleep can produce a dramatic reduction in their symptoms, although they return as soon as the deprivation is stopped .

If you take chronic, medicated schizophrenic patients, they have this huge deficit in what are called sleep spindles, which are these one-second long EEG oscillations. Chronic medicated patients have a 40 percent reduction in sleep spindles compared to controls; Giulio Tononi's group [at the University of Wisconsin] and our group have found the larger the reduction in spindles, the more positive symptoms [hallucinations and delusions, for instance] the patient shows. And the reduction in spindles correlates with a reduction in sleep-dependent memory enhancement. We just had a paper out in the journal *Sleep* that shows that if you give these schizophrenic patients, Lunesta (eszopiclone), you can increase their spindles and sleep-dependent memory processing now looks normal.

Stix: Are we at a point that we can make recommendations on how much sleep someone needs each night?

Stickgold: Asking how much sleep you need at night is like asking how asking how much food you need. The average person would say 2000 calories. But you also need a certain amount of protein and vitamins, for example. The calories are just to support energy demands of the body.

When you ask me how much sleep someone needs I ask for what? It might be that the amount of sleep needed to consolidate episodic memory is different than the amount of sleep you need to extract gist from information, and both may be different from what you need to optimize glucose regulation and antibody production.

If you take college students and put them on four hours of sleep a night for five days and then do a glucose tolerance test, they're looking pre-diabetic. That's not enough sleep for maintaining normal insulin regulation, but for some memory tests four hours may be more than enough.

And we have to ask how much sleep when? For some memory processes, a 90-minute nap gives you as much benefit as a whole night of sleep and in some cases more benefit than six hours of sleep or four hours of sleep at night. There's something unusually efficient about naps in terms of these memory processes.

On the other hand, if you feel like the amount of sleep you're getting is enough, if you don't sleep longer on weekends, if you don't think it would be a joke to not set an alarm clock, if you don't drink coffee within the first two or three hours of getting up, you're probably doing fine.

If you drink two cups of coffee to get going in the morning, you don't have enough sleep. If you sleep two hours later on weekends, you are not getting enough sleep. I think the amount varies from person to person. There isn't an absolute amount. I tell people to turn off the alarm clock for a week and see what happens. If you discover you're waking an hour and a half after you're supposed to be at work, you're probably not going to bed early enough. I still think eight hours looks like the best bet if I were to guess.

About the Author

Gary Stix is a senior editor at Scientific American.

Why Sleep-Deprived People Are More Selfish and Lonely

By Daisy Yuhas

How did you sleep last night? For many people, the answer is "not great." In the U.S., about a third of people get fewer than seven hours of shut-eye, which is the minimum recommendation. Most of us are familiar with the unpleasant results. We may feel groggy and grumpy, and everything gets just a bit harder to do. But sleeplessness also impairs parts of the brain that affect our social lives and abilities to relate to other people, according to research by neuroscientist Eti Ben Simon, a postdoctoral fellow at the University of California, Berkeley. People who sleep less are less likely to help others, for instance. Ben Simon spoke with *Mind Matters* editor Daisy Yuhas to discuss what the social consequences of bad sleep are and how good sleep can be leveraged to benefit well-being.

[An edited transcript of the interview follows.]

Yuhas: In past research, you and your colleagues have found that lack of sleep contributes to anxiety and difficulty managing emotions. In addition, you've recently found that poor sleep affects social interactions, regardless of mood. How does that work?

Simon: People are less interested in social interaction when they're sleep-deprived. For example, we designed a task where an experimenter and participant would face each other, and they would walk toward each other. The participant would decide when someone got too close, and we would measure that distance. Consistently, when people were sleep-deprived, they preferred others to be farther away.

Mood may play a part in the social consequences of sleep loss, but it's not the whole story. We've controlled for mood in studies of social behavior and seen that social withdrawal is not just an effect of mood.

We've also found that sleep deprivation reduces activity in what's known as the theory of mind network in the brain. These are areas that help us think about other people–what they might want, what they're like, and how they are similar to or different from ourselves.

Yuhas: How might those changes in the brain relate to our thinking?

Simon: In a recent study, we explored that by asking participants–either after they had a full night's sleep or after they had had no sleep–to think about different individuals while their brain activity was being recorded in an fMRI [functional magnetic resonance imaging] scanner. Then they answered a questionnaire that explored how likely they were to help others. We asked about daily acts, such as keeping the elevator door open, helping someone with their grocery bags, things like that.

We found that sleep-deprived people were significantly less likely to want to help others, and that correlated one to one with the impairment in their theory of mind network. The larger the impairment, the less they wanted to help other people. And it didn't matter if it was helping a friend, a family member or a stranger.

Yuhas: Total sleep deprivation is pretty extreme. What if people get at least some sleep?

Simon: In a second study, we used the same questionnaire, but this time, we tracked people's natural sleep across four nights and days. We didn't find any association with how much people slept in hours, but we found that the *quality* of their sleep–how many times they woke up during the night, whether they had a restful sleep or their sleep was fragmented–determined whether they were likely to want to help others or not.

Yuhas: So people who sleep poorly say they don't want to help others. But have you found that sleep deprivation changes actual helping behaviors in a noticeable way?

Simon: Yes. In a third study, we looked at a database of charitable donations and compared the week of [the transition to] daylight saving time—when people lose, on average, an hour of sleep—to other weeks of the year. In 15 years' worth of data, we kept seeing that on that week, the amount people donate is reduced by about 10 percent.

In Arizona and Hawaii, states that don't do daylight saving, we don't see this effect. We also looked at the switch to standard time, when people "fall back," and didn't find an effect either. Putting these data together, we could demonstrate that lack of sleep may be impairing altruistic sentiment across the U.S.

Other researchers are looking at similar questions. Just a few months ago a colleague of mine published a study that found doctors prescribe fewer painkillers during a night shift than a day shift, and that was accompanied by reduced empathy for their patients.

Yuhas: You've also found that one person's sleeplessness affects everyone around them. How does that happen?

Simon: In general, our ability to pick up that something's off with the person we're interacting with is really magnificent. In one study, just a one-minute video clip of someone talking was enough for others to pick up that they didn't want to interact with a sleep-deprived person.

Sleep deprivation can make people feel lonely, and we've found that when people come into contact with someone who is sleep deprived, they report feeling lonelier after that interaction. I like to say that helps settle the Beatles' question: "All the lonely people, where do they all come from?" It all started with sleep loss.

More seriously, I worry about a negative feedback loop with loneliness. If you're chronically sleep-deprived, that feeling of not connecting can just keep increasing. You're more withdrawn, less interested in interacting with others, and we've shown that others are less interested in interacting with you.

Yuhas: Could that in turn contribute to the connection between sleep and conditions such as anxiety or depression?

Simon: There is an intimate link between mental health and lack of sleep. We know that people suffering from disrupted sleep are also more likely to develop anxiety and depression down the line. With anxiety, for instance, when something stresses you out, the first thing that gets impaired is your sleep. And paradoxically, more sleep is what you need to feel calmer.

Yuhas: We're such social beings. Why would our sleepy brain shut down the network that helps us connect with others?

Simon: The brain and body need sleep so much that they start letting go of whatever we don't immediately need once sleep is missing. In nature, the only time you see animals sleep-deprived is when they're migrating, they just had a baby or they're starving. So being sleep-deprived is really a stress signal that makes us want to accumulate as much food as we can and be more alert to threats. We don't have the capacity for anything more than that.

The more time you spend awake, the more sleep becomes the one thing that you need to focus on. Everything else just goes in the background.

Yuhas: On the flip side, does this mean we should really put better sleep alongside diet and exercise as a crucial way to improve mental health?

Simon: Exactly. Sleep is something that can help us intervene into social and emotional affective disorders; it's another way into mental health.

Sleep is actually the bedrock, even before diet and exercise. The way we metabolize food and the way our muscles react also depends on good sleep.

It's not always easy to restore good sleep. But once we prioritize it, we can really help shift symptoms, mood, anxiety and even the desire to be kind to other people.

About the Author

Daisy Yuhas edits the Scientific American *column* Mind Matters. *She is a freelance science journalist and editor based in Austin, Texas. Follow Yuhas on Twitter @DaisyYuhas.*

Sleep Deprivation Shuts Down Production of Essential Brain Proteins

By Emily Willingham

Most of us could use more sleep. We feel it in our urge for an extra cup of coffee and in a slipping cognitive grasp as a busy day grinds on. And sleep has been strongly tied to our thinking, sharpening it when we get enough and blunting it when we get too little.

What produces these effects are familiar to neuroscientists: external light and dark signals that help set our daily, or circadian, rhythms, "clock" genes that act as internal timekeepers, and neurons that signal to one another through connections called synapses. But how these factors interact to freshen a brain once we do sleep has remained enigmatic.

Findings published on October 10 in two papers in *Science* place synapses at center stage. These nodes of neuronal communication, researchers show, are where internal preparations for sleep and the effects of our sleep-related behaviors converge. Cellular timekeepers rhythmically prep areas around the synapses in anticipation of building synaptic proteins during slumber. But the new findings indicate neurons curtail their regular on-off cycles of building these critical proteins in the absence of sleep.

The results suggest the brain is "getting prepared for an event, but it doesn't mean you actually follow through on doing it," says Robert Greene, a neuroscientist at the University of Texas Southwestern Medical Center, who was not involved in the study. Greene calls the studies "fascinating," saying they confirm a "long suspected" connection between internal timekeeping and sleep behaviors.

When we become sleepy, two factors are in play: "sleep pressure," or the growing allure of a beckoning pillow as waking time lengthens, and our internal clock sounding the signal that the usual point for shut-eye has arrived. In one of the two studies, Sara B. Noya of the Institute of Pharmacology and Toxicology at the University of Zurich and her colleagues showed that in mice, the internal clock regulates the rhythmic generation of instructions, or transcripts, for making proteins. Giving in to sleep pressure and hitting the hay, they found, triggers the final steps of protein production.

At two peak times in the 24-hour day, just before waking and sleeping, neurons in cognition-related brain areas packed a timekeeping cell's signaling stations with these transcripts, Noya's team discovered. The "sleep time" transcripts tended to be for proteins that regulate building other proteins, while the "wake time" instructions were for proteins linked to synapse function. These stashed molecules set the stage for the rapid refreshing of synapses during sleep. Mice lacking important clock genes did not show these peaks.

With a regular sleep-wake cycle, the proteins built using these instructions also showed peak production at dawn and dusk. In sleep-deprived mice, however, Noya and her colleagues demonstrated that the cell still produced many of the transcripts but did not build the related proteins. That result implies sleeping regulates the final, protein-building step in ensuring robust synapses.

Not all proteins that the cell makes necessarily go into active service, though. In a companion paper, Franziska Brüning of the Ludwig Maximilian University of Munich and the Max Planck Institute of Biochemistry in Martinsried, Germany, and her colleagues explored the rhythmic use of those that do. Attachment or removal of a phosphate molecule acts as a toggle to turn proteins on or off, so the investigators took a close look at this process. They found levels of proteins that had been tagged with phosphates also peaked twice, with the bigger peak occurring just before waking. And as with proteins in the other study, sleep deprivation flattened these peaks.

The researchers made their measurements every four hours, an advance on earlier studies that usually looked at a single time point during a 24-hour period, says Chiara Cirelli, a neuroscientist at the University of Wisconsin–Madison, who co-wrote a commentary accompanying the two papers. "It's a very comprehensive analysis across the entire light-dark cycle," she says.

Cirelli emphasizes the importance of isolating the synaptic regions where these molecules accumulate and are produced. The researchers pinned down when transcripts were positioned at the ready and when proteins—tagged with phosphates or not—were made or used, she says.

Maria Robles, a neuroscientist at Ludwig Maximilian University of Munich and a co-author of both papers, says the findings distinguishing the different stages of protein production and activity are eye-opening, revealing the brain has "a beautiful way to control" these molecules.

Even though the studies were done in mice, the brains of these animals have proved to be pretty reliable substitutes for those of humans, says Akhilesh B. Reddy, a neuroscientist at the Perelman School of Medicine at the University of Pennsylvania, who was not involved in the work. The findings have implications for how we consolidate memories during sleep, among other promising avenues of research, he says, drawing the focus straight to events at the synapse. That does not mean, however, that interventions to boost memory and cognition loom in the near future, based on these findings, Robles says. "This is just the tip of the iceberg," she adds.

About the Author

Emily Willingham is a science writer and author based in Texas. Her latest book is The Tailored Brain *(Basic Books, 2021).*

Acting Out Dreams Predicts Parkinson's and Other Brain Diseases

By Diana Kwon

Alan Alda was running for his life. The actor, best known for his role on the television series *M*A*S*H*, wasn't on a set. This threat was real—or at least it felt that way. So when he saw a bag of potatoes in front of him, he grabbed it and threw it at his attacker. Suddenly, the scene shifted. He was in his bedroom, having lurched out of sleep, and the sack of potatoes was a pillow he'd just chucked at his wife.

Acting out dreams marks a disorder that occurs during the rapid eye movement (REM) phase of sleep. Called RBD, for REM sleep behavior disorder, it affects an estimated 0.5 to 1.25 percent of the general population and is more commonly reported in older adults, particularly men. Apart from being hazardous to dreamers and their partners, RBD may foreshadow neurodegenerative disease, primarily synucleinopathies—conditions in which the protein α-synuclein (or alpha-synuclein) forms toxic clumps in the brain.

Not all nocturnal behaviors are RBD. Sleepwalking and sleep talking, which occur more often during childhood and adolescence, take place during non-REM sleep. This difference is clearly distinguishable in a sleep laboratory, where clinicians can monitor stages of sleep to see when a person moves. Nor is RBD always associated with a synucleinopathy: it can also be triggered by certain drugs such as antidepressants or caused by other underlying conditions such as narcolepsy or a brain stem tumor.

When RBD occurs in the absence of these alternative explanations, the chance of future disease is high. Some epidemiological studies suggest that enacted dreaming predicts a more than 80 percent chance of developing a neurodegenerative disease within the patient's lifetime. It may also be the first sign of neurodegenerative disease, which on average shows up within 10 to 15 years after onset of the dream disorder.

One of the most common RBD-linked ailments is Parkinson's disease, characterized mainly by progressive loss of motor control. Another is Lewy body dementia, in which small clusters of α-synuclein called Lewy bodies build up in the brain, disrupting movement and cognition. A third type of synucleinopathy, multiple system atrophy, interferes with both movement and involuntary functions such as digestion. RBD is one of the strongest harbingers of future synucleinopathy, more predictive than other early markers such as chronic constipation and a diminished sense of smell.

Descriptions of dream enactment by people with Parkinson's are as old as recognition of the disease itself. In James Parkinson's original description, "An Essay on the Shaking Palsy," published in 1817, he wrote: "Tremulous motions of the limbs occur during sleep, and augment until they awaken the patient, and frequently with much agitation and alarm." But despite similar reports over the next two centuries, the connection between dreams and disease remained obscure—so much so that Alda had to convince his neurologist to do a brain scan for Parkinson's after he read about the link in a 2015 news article.

Those scans confirmed Alda's suspicion: he had Parkinson's. He shared his experience with the public "because I thought anybody who has any symptom, even if it's not one of the usual ones, could get a head start on dealing with the progressive nature of the disease," he says. "The sooner you attack it, I think, the better chance you have to hold off the symptoms."

In recent years awareness of RBD and an understanding of how it relates to synucleinopathies have grown. Studying this link is giving researchers ideas for early intervention. These advances contribute to a growing appreciation of the so-called prodromal phase of Parkinson's and other neurodegenerative disorders—when preliminary signs appear, but a definitive diagnosis has not yet been made. Among the early clues for Parkinson's, "RBD is special," says Daniella Berg, a neurologist at the University

Hospital Schleswig-Holstein in Germany. "It's the strongest clinical prodromal marker we have."

Lifting the Brake

Ray Merrell, a 66-year-old living in New Jersey, started acting out his dreams around 15 years ago. His dreamscapes became action-packed, like "something you'd watch on TV," Merrell says. He often found himself either being chased by or chasing a person, animal or something else. In the real world, Merrell was flailing, kicking and jumping out of bed. Some of his violent nighttime behaviors injured him or his wife.

In people with RBD, the brakes that normally immobilize them during REM sleep–the stage of sleep most closely linked with dreaming–are lifted. (Dreaming also occurs in non-REM sleep, but dreams during REM are longer, more vivid and more bizarre.)

In the 1950s and 1960s French neuroscientist Michel Jouvet conducted a series of experiments that revealed just how chaotic unrestricted movements during REM sleep could be. By lesioning parts of the brain stem in cats, Jouvet inhibited the muscle paralysis that occurs in many species during REM sleep. Cats that had gone through the procedure acted normally when awake, but when asleep they became unusually active, exhibiting intermittent bursts of activity such as prowling, swatting, biting, playing and grooming. Despite this remarkably awakelike behavior, the cats remained fast asleep. Jouvet observed that the cats' sleeping actions often were unlike their waking habits. Felines that were "always very friendly when awake," he wrote, behaved aggressively during REM sleep.

In the late 1980s Carlos Schenck, a psychiatrist at the University of Minnesota, and his colleagues published the first case reports of RBD. Patients described having violent dreams and aggressive sleep behaviors that contrasted sharply with their nonviolent nature while awake–echoing Jouvet's documentation of otherwise friendly felines that turned belligerent during sleep. One patient, for example, said he had a dream about a motorcyclist trying to ram him on the highway. He turned to kick the bike away–and woke to his wife saying, "What in heavens are you doing to me?" because he was "kicking the hell out of her." Another said he dreamed of breaking a deer's neck and woke up with his arms wrapped around his spouse's head.

To test whether these bizarre behaviors may reflect damage to the brain stem, as in Jouvet's cats, Schenck and his colleagues kept track of such patients to see whether they might develop a brain disease. In 1996 they reported that in a group of 29 RBD patients, all of whom were male and age 50 or older, 11 had developed neurodegenerative disease an average of 13 years after the onset of their RBD. By 2013, 21 of them, or more than 80 percent, had developed a neurodegenerative condition–the most common of which was Parkinson's.

Subsequent studies confirmed this link. Of 1,280 patients across 24 centers around the world, 74 percent of people with RBD were diagnosed with a neurodegenerative disease within 12 years. Sometimes RBD shows up decades before other neurological symptoms, although the average lag appears to be about 10 years. When dream enactment occurs alongside other early signs of synucleinopathies, people tend to develop a neurodegenerative disease more rapidly.

Many researchers expressed skepticism about this link early on, says Bradley Boeve, a professor of neurology at the Mayo Clinic in Rochester, Minn. "We would get reviewer comments back saying that this is hogwash," he says. But the connection between RBD and synucleinopathy has become well accepted: "I think that's pretty much gospel now."

Some scientists suspect RBD results from an aggregation of synuclein and associated neurodegeneration in areas of the brain stem that immobilize us during REM sleep. In its normal, benign, form, the protein is involved in the functioning of neurons, but when "misfolded" into an atypical configuration, it can form toxic clumps. Autopsies have shown that more than 90 percent of people with RBD die with signs of synuclein buildup in their brains. There are no established methods to probe for synuclein clusters in the brains of living patients, but scientists have looked for the toxin in other parts of the body. Alejandro Iranzo, a neurologist at the Hospital Clinic Barcelona in Spain, and his colleagues were able to detect misfolded synuclein in the cerebrospinal fluid of 90 percent of patients with RBD.

As an early manifestation of Parkinson's and related diseases, RBD can help scientists trace the ways in which toxic synuclein spreads throughout the body and brain. Evidence is mounting that at least in some patients, pathology may begin in the gut and spread up through lower brain structures such as the brain stem to the higher regions influencing movement and cognition. One likely pathway is the vagus nerve, a bundle of nerve fibers connecting all the major organs with the brain. Alpha-synuclein clumps injected into the guts of mice can spread to the brain via the vagus—and in humans, at least one epidemiological study has shown that cutting the vagus, a procedure sometimes used to treat chronic stomach ulcers, decreases the risk for Parkinson's later in life.

Some researchers hold that Parkinson's has two subtypes: gut first and brain first. RBD is highly predictive of later Parkinson's, says Per Borghammer, a professor of clinical medicine at Aarhus University in Denmark, but the converse is not true: only about a third of people with Parkinson's get RBD before developing motor symptoms. People with RBD have gut-first Parkinson's, Borghammer posits, and generally experience symptoms such as constipation long before motor and cognitive decline. But in the two thirds of patients who are brain first, RBD may emerge later than problems with movement—or never appear.

The Dream Theater

Does damage to the brain stem also affect the content of dreams and the actions of dreamers? Sleep researcher Isabelle Arnulf, a professor of neurology at Sorbonne University in Paris, developed a keen interest in the dream-time behaviors of her Parkinson's patients after noticing an unusual pattern: although these people struggled with movement while awake, their spouses often reported that they had no trouble moving while asleep. One particularly memorable patient, according to Arnulf, had been dreaming of crocodiles in the sleep lab when he lifted a heavy bedside table above his head and loudly shouted, "Crocodile! Crocodile!" to an empty room. When awake, he struggled to lift objects and to speak.

Intrigued by such observations, Arnulf and her colleagues began compiling the behaviors people exhibited during REM sleep. This collection, which has grown over the past decade and a half to include hundreds of hours of footage of dream-enacting sleepers and hundreds of dream reports, has enabled Arnulf to uncover unexpected features of RBD dreams and insights into some fundamental questions about how—and why—we dream.

Merrell, Alda and many other people with RBD often have dreams in which they face danger. In one study led by Arnulf, researchers found that among people with RBD, 60 percent reported dreams involving some kind of threat, and 75 percent confronted their attacker instead of running away. People who report more frequent distressing dreams are also at greater risk of developing Parkinson's. "It's textbook for people with RBD to have violent dreams where they are on the defensive," says Yo-El Ju, a professor of neurology at Washington University School of Medicine in St. Louis. But whether this is attributable to recall bias—people tending to remember more violent dreams because they are more memorable—remains an open question, she adds.

Arnulf's group also found that a significant proportion of RBD dreams are nonviolent. In one study, 18 percent of patients flew, sang, danced, laughed, lectured or enacted other peaceable activities. In another study with 52 RBD patients, the researchers looked at subtle changes in facial expressions during sleep. Half the people smiled and a third laughed during mainly REM sleep, suggesting that RBD dreams may be more positive than previously described. Arnulf hypothesizes that violent dreams may be reported more often because aggressive behaviors are more likely to wake up the dreamer or their spouse. "I'm pretty convinced that in RBD patients, it's just that the window is open on dreaming, but their dreams are not different from ours," Arnulf says.

The finding that RBD patients display a range of emotions while dreaming led Arnulf to believe that what researchers learn about their dreams may apply to the broader population. Her team discovered, for example, that a small percentage of people with RBD were never able to recall their dreams despite acting out dreamlike behaviors while asleep—suggesting that self-described nondreamers may, in fact, dream.

One mystery of RBD is whether people are acting out their dreams or whether their movements are modifying their dream narratives, says Birgit Högl, a professor of neurology and sleep medicine at the Medical University of Innsbruck in Austria. As for the question that originally intrigued Arnulf—why the impaired movement characteristic of Parkinson's seems to disappear during sleep in some patients—work by other groups has helped suggest an answer. Neurologist and psychiatrist Geert Mayer, formerly at Hephata Clinic in Germany, and his colleagues revealed in a 2015 study that the basal ganglia, movement-related structures near the base of the brain where neurodegeneration occurs in people with Parkinson's, were silent during dream enactments in RBD patients. But other brain regions involved in producing movement, such as the motor cortex, were active.

Findings such as these suggest that in people with RBD, movement is generated through a motor circuit that bypasses the basal ganglia. "This sort of shows that whatever's going on in Parkinson's disease in terms of your movement doesn't apply to you when you're asleep," says Ronald Postuma, a professor of neurology at McGill University. It also raises a tantalizing possibility for therapy: "What if you could mimic whatever that motor state is when a person is asleep but keep them otherwise awake?"

Early Intervention

Merrell had been enacting dreams for several years before he realized it might be a sign of a bigger problem. It began during a rough patch at work, and he had dismissed the occasional sleep outbursts as deriving from job-related stress. One evening, mid-dream, Merrell threw himself into a corner of a nightstand, breaking his skin but narrowly missing his breastbone. The close call with a very serious injury "really got me thinking that I better look into this," he says.

When Merrell was diagnosed with RBD in 2011, his doctor briefly mentioned the risk of developing other conditions down the line but "didn't give me assurances or any other advice," Merrell recalls. But when he began researching the condition online, he discovered many studies on RBD patients who developed a neurodegenerative disease in later life. "The more I searched," he adds, "the more I realized, wow, this has some pretty significant implications."

The available treatments for Parkinson's and other synucleinopathies can currently only manage symptoms. They're unable to slow or stop the underlying neurodegeneration. "The worst news I have to give as a sleep doctor is to tell someone that they have RBD," Ju says.

But several new therapeutics for Parkinson's and other synucleinopathies are being developed, and many neurologists believe early intervention could be crucial. "The Parkinson's disease field, in particular, is full of failed treatment trials," Ju says. "By the time people have the disease, it's probably too late to intervene–too many cells have died." Going back and testing these seemingly failed medications in RBD patients may prove more successful, she adds, because as a much earlier stage of disease, RBD provides a window where treatments are more likely to be effective. "A lot of people are viewing RBD as similar to high cholesterol," Boeve says. "If you have high lipid levels, they increase your risk for heart disease and stroke. If you can alter that pathophysiological process, you can reduce the risk or delay the onset."

Ju, Postuma and Boeve are co-leaders of the North American Prodromal Synucleinopathy (NAPS) Consortium, which launched in 2018. The NAPS investigators aim to pinpoint clinical and biological markers through various means, including brain scans, genetic screens, and tests of blood and cerebrospinal fluid. The researchers hope these markers will eventually indicate how and when a person with RBD will develop a neurodegenerative disease later in life—and which disease they will end up with. Ideally, such biomarkers would help scientists identify RBD patients for investigative therapies that target α-synuclein years before debilitating symptoms appear. The ultimate goal of NAPS, Ju says, "is essentially to prepare for clinical trials for protective treatments."

In 2021 NAPS received a $35-million grant from the National Institutes of Health for this work, which will be carried out across eight sites in the U.S. and one in Canada. In a parallel effort, Högl, along with other researchers in Europe, is gathering a similar cohort of patients from multiple institutions across the continent for future clinical studies. Wolfgang Oertel, a neurologist at Philipps University of Marburg in Germany, who is involved in the European effort, is optimistic about the future for people with RBD. He expects that of the dozens of potentially disease-modifying Parkinson's drugs currently in clinical trials, at least a few will be available soon. "I tell my patients, 'You've come at the right moment,'" Oertel says. "You will be one of the first to get the right drugs."

Högl has also been involved in another active area of investigation: finding ways to better characterize RBD. Working with Ambra Stefani of the Medical University of Innsbruck and other colleagues, she has been gathering measurements of muscle activity during sleep in people with RBD. They hope that this work will not only help to streamline the diagnosis of RBD but also help doctors to detect the sleep disorder even earlier, in so-called prodromal RBD, where overt dream enactments might not occur, or in people who may have RBD but exhibit only small, difficult-to-detect movements. Their work suggests that the elaborate, violent behaviors seen in RBD are "just the tip of the iceberg," Högl says. They may occur on one night but not another. Minor muscle jerks in the hands or elsewhere, in contrast, appear to be much more frequent–and a more stable sign because they occur hundreds of times during the night, she adds.

For now there is no cure for RBD or Parkinson's–but that doesn't mean there is nothing patients can do. A growing body of evidence indicates that moderate to intense exercise helps to improve both motor and cognitive symptoms of Parkinson's, and many neurologists already recommend such physical activity to their patients with RBD. "The evidence suggests that the benefits of exercise are more than just symptomatic," says Michael Howell, a neurologist at the University of Minnesota. "It appears that this actually is helping to protect brain cells."

Both Alda and Merrell have taken that advice to heart. In addition to medications, Alda has taken up exercise-based therapy for Parkinson's. Merrell, too, has integrated regular physical activity into his routine, hiking for several miles every other day. He's gotten involved in clinical research and is one of the NAPS participants. This contribution helps Merrell feel empowered–he hopes to aid the discovery of effective neuroprotective therapeutics. "Somebody always had stepped up in other illnesses or conditions that allowed for clinical trials and the therapies that we have today," Merrell says. "I just happened to be queued up for this–and I accept that challenge."

About the Author

Diana Kwon is a freelance journalist who covers health and the life sciences. She is based in Berlin.

Sleep Apnea Is Different for Women

By Claudia Wallis

Picture, if you will, your typical sleep apnea sufferer. Chances are he is middle-aged and overweight and snores like a freight train. Note the male pronoun. Twenty-five years ago experts believed that the condition, in which breathing is disrupted during sleep, was about 10 times as common in men as in women. Better-quality studies have since reduced that ratio to roughly three to one, but as more data come to light, it is becoming clearer that sleep apnea—and the broader category known as sleep disordered breathing—simply looks a little different in women. And that suggests it is often overlooked.

Sleep apnea is a concern because it raises the risk of heart attacks, hypertension, arrhythmias, insulin resistance, strokes and accidents that result from daytime sleepiness. Put simply, gasping for breath at night and not giving your body a thorough rest puts a lot of pressure on the cardiovascular system, raises adrenaline levels and ignites inflammation. Doctors diagnose apnea with a sleep test, often done at home, that measures your apnea-hypopnea index. This index reflects the average number of times an hour that you have an episode lasting at least 10 seconds during which breathing stops (apnea) or becomes so deficient that blood oxygen levels fall by 3 or 4 percent or more (hypopnea). Fewer than five such episodes an hour is considered normal. Five to 15 is mild sleep apnea, 15 to 30 is moderate and more than 30 is severe.

Most home tests do not, however, examine the *stage* of sleep in which these episodes occur, and that may be a problem. A growing body of evidence shows that for many women disrupted breathing is concentrated in the rapid eye movement (REM) phase, which is also when dreams are most vivid and when heart and respiratory rates become less regular. A study of 2,057 men and women aged 45 to 84, published last November in the journal *Sleep*, found that women have just as many events as men do during the REM phase.

"Whatever protection women have in non-REM sleep is not there during REM sleep," says Christine Won, medical director of the Yale Centers for Sleep Medicine and lead study author. Disrupting REM may be especially bad for health. "Studies suggest that how many events you have during REM sleep is what really puts you at risk for cardiovascular health effects," Won says. But because REM accounts for only about 20 percent of a person's nightly slumber, a test that averages events across the entire night can be misleading.

Several other sex differences emerged from the new study. "One of our findings is that women have a lower arousal threshold–they are more likely to wake up at night in response to a given apnea," says Susan Redline, senior author of the study and a senior physician at the Division of Sleep and Circadian Disorders at Brigham and Women's Hospital in Boston. This may also impact diagnosis. Women are more likely than men to briefly wake up before their oxygen level falls by more than 4 percent–the threshold used to identify and treat sleep apnea in patients covered by Medicare. When Won and Redline used a cutoff of 3 percent oxygen desaturation, many more women met the criteria for sleep apnea.

These differing patterns may help explain why women with sleep apnea are more likely to complain about morning headaches, fatigue, depressed mood and insomnia. In men, a big complaint (usually from a bed partner) is loud snoring, along with daytime sleepiness. In both sexes, apnea rates rise with obesity and age.

Hormones most likely play a role in these sex differences. Sleep apnea increases in women after menopause, and it is common in women with polycystic ovary syndrome, a condition characterized by high levels of testosterone. Redline notes that anatomy may also be a factor: "Men have a longer and more collapsible airway." As for the female tendency to wake up more easily, evolutionary pressures and experience might be at work: "It may be that women evolved to wake up to tend to their children," Redline says.

The leading therapy for apnea is the use of a CPAP (continuous positive airway pressure) machine to force air into the throat, keeping the airway open. Research suggests that it works equally well for REM and non-REM apnea. But as scientists shed more light on the varying patterns of the condition, treatment may become more tailored to the individual. Diagnostic criteria might also need to change to capture more cases in women. Medicare's 4 percent desaturation threshold is one example. And the increasing use of at-home testing rather than costlier testing in a sleep lab may be another. "If our findings are true," Won says, "then home sleep apnea testing biases against diagnosing women."

About the Author

Claudia Wallis is an award-winning science journalist whose work has appeared in the New York Times, Time, Fortune *and the* New Republic. *She was science editor at* Time *and managing editor of* Scientific American Mind.

Sleep Deprivation Sometimes Relieves Depression. A New Study May Show Why

By Emily Willingham

In 1818 Johann Christian August Heinroth, considered to have been the first professor of psychiatry at a university, suggested that sleep deprivation might alleviate "melancholia," or depression. But it wasn't until 1959 that formal reports began to emerge, again from Germany, suggesting that a night of sleeplessness could boost mood in depression. Experimental trials in the 1970s went on to confirm a benefit. Since then study after study has shown that spending a night without sleep, especially with lights on, indeed produces mood benefits for about half of the people with depression.

The effects of this approach, dubbed "wake therapy," offer the bonus of being immediate, unlike most antidepressants, which require a few weeks to work. Sleep deprivation has disadvantages that include, well, going without sleep. As anyone who has parented an infant can attest, that has unwanted "spillover" effects on other aspects of life. Identifying processes in the brain that underlie sleep-deprived boosting of mood could lead to therapies that are less burdensome than enduring a wakeful night.

A new study published June 20 in the *Proceedings of the National Academy of Sciences USA* has identified specific brain regions that kick up activity when sleep deprivation lifts one's mood. Because the research included people with and without depression, the findings broaden understanding about the "bizarre phenomenon" of mood-boosting sleep deprivation, says study author Philip Gehrman, a professor of clinical psychology at the University of Pennsylvania.

Even without immediately leading to new therapies, the results confirm the benefit of wake therapy in depression, says Anna Wirz-Justice, a professor emerita at the Center for Chronobiology at the University Psychiatric Clinics Basel in Switzerland, who was not involved in the work. "Perhaps this study, which provides clues to mechanisms, will lead to reevaluation of the intervention as an inexpensive, rapid antidepressant modality."

To take a peek inside the brain, Gehrman and his colleagues evaluated 30 people with major depressive disorder who all underwent sleep deprivation. They also assessed another 54 people without depression, 16 of whom served as controls who did not undergo sleep deprivation.

Over the course of five days, researchers performed three imaging scans on all of the participants. The sleep deprivation groups underwent one scan after a normal night's sleep, another after a sleepless night and a third after two nights of recovery sleep. The 16 control participants without depression underwent three scans, too, but got regular sleep. The scanning tracked blood oxygen delivery in the brains of participants as they lay still, doing nothing. In between scans, every two waking hours from days two to five, all participants completed a questionnaire assessing their mood.

The questionnaire responses showed that 43 percent of those with diagnosed depression experienced mood improvement after sleep deprivation. Most, but not all, of those without depression reported a worsened mood after losing sleep.

Imaging in those reporting improved mood showed increased activity in two brain regions that were previously implicated in depression and the effects of sleep deprivation. One of these regions, the amygdala, is famously associated with emotion processing and memory. The other is the anterior cingulate cortex, which studies have connected to depression and to the benefits of sleep deprivation. Unexpectedly, activity in these two areas was increased in participants whose mood improved whether they had depression or not.

The involvement of the anterior cingulate cortex "fits the picture" that these previous studies have pieced together about depression and disrupted sleep, says Francesco Benedetti, a longtime sleep disruption researcher at the department of clinical neurosciences at San Raffaele Hospital in Milan, who was not involved in the work.

The implied link between the two regions suggests that they have a role in producing an elevated mood after a sleepless night. In participants with depression, the connection persisted even after two nights of recovery sleep. Benedetti says that sleep deprivation may heighten the effect of structures in the upper part of the brain, such as the anterior cingulate cortex, on quieting the amygdala and boosting mood. This signal in people reporting mood improvement regardless of depression diagnosis is "an important clue" to understanding how sleep deprivation works, Wirz-Justice says. Studies like this are a "fabulous" way to study what underlies changes in mood without interference from medications, she adds.

Back in 1976 Burkhard Pflug, a psychiatrist at the University of Tubingen in Germany, wrote that sleep deprivation might behave like a "zeitgeber," or "time giver," in people with depression and resynchronize aberrant brain rhythms. Benedetti's work indicates that depression might flatten rhythmic cycles that support brain function and that sleep deprivation might jump-start these rhythms back into a healthy pattern.

The new findings may offer targets for reviving the zeitgeber in depression in accessible ways. A noninvasive technique called transcranial magnetic stimulation, in which magnetic waves are applied from outside the skull, is used to treat depression, but its effectiveness relies on hitting the right regions. Gehrman says that brain circuits that are responsive to sleep deprivation could represent candidate targets for transcranial magnetic stimulation or other kinds of stimulation. Such approaches hold potential to produce the rapid effect of sleep deprivation without the downside of a sleepless night.

About the author

Emily Willingham is a science writer and author based in Texas. Her latest book is The Tailored Brain *(Basic Books, 2021).*

Section 4: The Importance of Good Rest for Kids and Teens

Why Rocking to Sleep Is a Matchless Sedative—and Elixir

By Bret Stetka

Rocking babies to sleep—to both quiet the wails of youth and preserve the sanity of young parents—has been commonplace dating back to prehistory. Similarly, rhythmic motions like the muted clank of a train ride coax many of us adults into an instant slumber—but why?

Two new studies published today in *Current Biology* suggest our brains are evolutionarily programmed to respond to rocking. The research shows in both humans and mice, rocking to sleep may have significant health benefits such as better quality of sleep and even improved long-term memory formation.

In the first study the authors used electroencephalography (EEG) recordings to analyze the brain responses of 18 young healthy adults who spent three nights in a sleep lab. The first night was intended to acclimate them to their new sleep environment; the second they spent in a slowly rocking bed; the third—one week later—in a stationary bed.

Even in people who were already good sleepers, rocking shortened the time it took for subjects to fall asleep, and also to reach non-REM sleep, which correlates with improved sleep quality. The rockers also had fewer arousals and maintained deep sleep for a longer period of time.

Curious if rocking might have any influence on memory, study subjects were also tasked with memorizing 46 random word pairs. On both their rocking and non-rocking sleepovers they were then presented with the first word and asked to recall the second after a night's sleep. A night of rocking improved their recall threefold.

Finally rocking also was found to synchronize brain waves during non-REM sleep in the thalamocortical networks of the brain, a system known to be involved in both sleep and memory consolidation, the nocturnal process during which the brain processes and stores long-term memories. "What was surprising...is that we clearly showed that specific brain oscillations of non-REM sleep are synchronized and entrained by the rhythm of the bed motion," says neuroscientist Laurence Bayer of the University of Geneva in Switzerland, who led the new study. Bayer likens the findings to our intuitive appreciation that a slowly rocking hammock can relax us and do wonders for our mood. "This is potentially a very promising avenue for future clinical research," he notes.

Previous research–and millennia of relieved parents–support the fact rocking helps coax infants to sleep. And the new findings suggest the same is true in adults. Yet it was previously unknown whether or not the same is true in other species.

The second study found that, at least in mice, it is. Like in the human study, rocking shortened the time it took the mice to fall asleep and increased the amount of sleep time, as measured by EEG recordings. Unlike humans, however, rocking did not seem to induce deeper sleep in the rodents. The authors suspected rocking might exert its influence on sleep by stimulating the vestibular system, the sensory organs in the inner ears of mammals that control our sense of balance and spatial orientation. Using a strain of mice with impaired vestibular function, they showed this is indeed the case. Specifically, half of the mice used in the study lacked "otoliths," small calcium carbonate–based particles that bob around two compartments of the inner ear and allow mammals to perceive vertical and horizontal acceleration. Mice without these otolithic organs showed none of benefits of rocking during sleep. "Since immemorial times we've rocked our babies, and often ourselves, to sleep," says lead author on the new paper, Konstantinos Kompotis, at the University of Lausanne, also in Switzerland. "Our study provides new insights into the neurophysiological mechanisms underlying this phenomenon."

"These findings are critically important," adds Bryce Mander, an assistant professor of psychiatry and human behavior at the University of California, Irvine, who was not involved in these studies. "Before these papers, both the mechanisms and the functional significance of the association between vestibular-related rocking and sleep were completely unknown." He notes the new findings open the door to testing non-invasive behavioral interventions in those with impaired sleep and memory.

Kompotis says the two studies could lead to improved treatments for people who have trouble falling asleep as well as for those with inner ear defects. "Rocking cradles for babies; rocking beds for younger adults; rocking chairs for the elderly–there are ongoing efforts out there that could improve our sleep and well-being," Kompotis adds. But for those hankering to add rocking to their fitness routines, he urges patience. "We should understand the phenomenon in its entirety before we start hanging hammocks in our–and our pets'–homes."

About the Author

Bret Stetka is a writer based in New York City and editorial director of Medscape Neurology (a subsidiary of WebMD). His work has appeared in Wired, *NPR and the* Atlantic. *He graduated from the University of Virginia School of Medicine in 2005.*

Let Teenagers Sleep

By The Editors

Teenagers are some of the most sleep-deprived people in the U.S. On average, teens do not get enough sleep, and more important, they do not get enough quality sleep, researchers say. We could blame cell phones and other light-emitting technologies for keeping kids up at night, but late nights are just part of the equation. In addition to technology, one fairly indisputable factor contributes to this collective sleepiness: school start times.

Over decades researchers have amassed evidence showing that pushing back the first bell of middle and high school would benefit the physical, mental and emotional health of older children, not to mention their academic performance. The Centers for Disease Control and Prevention, along with several medical societies, has endorsed later start times. Some school districts, as well as the state of California, have already shown respect for that evidence with new start times.

Yet far too many school districts are reluctant to make the change, whether for logistical, financial or cultural reasons. This is unfair to teens. A generation of students is playing catchup from COVID, and we need to prioritize their health and wellness by pushing back the start of the school day. Honoring their biological and social needs will create more resilient adults who can thrive in a world filled with current complexities and future ones we can't begin to predict.

Teenagers need about nine hours of sleep a night—but they get closer to seven. And around puberty, their circadian clocks shift by a couple of hours, meaning they get tired later at night than before and wake up later in the morning than they used to. This shift reverses at adulthood. The biological nature of this daily rhythm means that sending a teenager to bed earlier won't necessarily mean they fall asleep earlier.

Experts tell us that teens are missing out on both restorative sleep and REM sleep, especially the cycles that normally happen just before a person wakes up. Restorative sleep helps to repair the body after a hard day, and it may improve immune function and other biological processes. REM sleep solidifies events and learning into memories. So when a 10th grader who naturally goes to bed around 11 p.m. has to wake up at 6 a.m. for school, that teen is losing not only hours of sleep but hours of quality sleep. And even if they sleep in on the weekends, they won't fully catch up.

These kids are telling us they need more sleep. In survey after survey, they say when school starts later, they are not as tired all day, they tend to get to school on time, and they are less likely to have to be nagged to get out of bed. They tell us that as their sleep time decreases, their use of tobacco and drugs increases, including drugs that could help them stay awake. They tell us that getting one less hour of sleep a day leaves them feeling hopeless and, sometimes, suicidal. Research has shown that suicide risk in children increases during the school year, and sleep deprivation could be a contributing factor. Other studies show that getting one less hour of sleep a day is associated with weight gain. Researchers have told us that sleepy teens are more prone to car crashes and that even 30 minutes of extra sleep would help alleviate some mental health concerns. Even teachers have reported that with later start times, their students are more engaged in the morning, and teachers themselves are more rested.

Despite decades of research, thousands of publications and clear science, schools in only a few states and the District of Columbia have pushed their start times to 8:30 a.m. on average, which researchers say is a compromise—a better time would be closer to 9 a.m.

The path to delayed school start times is riddled with potholes. Bus schedules have to change. Teacher and administrator schedules have to be altered. After-school sports and enrichment programs might have to begin later. Parents and caretakers with more than one child may have to juggle child care for older children to get the younger ones to their earlier start times. A delayed school start could also mean adults with inflexible work schedules are late for work.

Experts say our agrarian model of education was designed to get teens up early and home before dark to tend to the farm, but it is no longer relevant for most modern students. Our cultural views of teens as lazy and of needing sleep as a weakness are harmful and inaccurate. And our grumbling that if *we* survived early start times, today's teens can, too, is callous and dismissive of science.

Access to education is a basic right in the U.S. But it's time to stop thinking of school start times as immovable mountains. While more states ponder start time legislation, school district administrators should prioritize it, and people running for school boards need to add start times to their platforms. State-level funding agencies have to clear hurdles for districts wanting to try this. Employers need to be more flexible to help parents adjust to school schedules, especially with hourly employees. And the unions that represent teachers and other education professionals need to negotiate with teens also in mind.

For decades we've ignored the overwhelming evidence that delayed start times help teens succeed. Let teenagers sleep. There is nothing "woke" about that.

High School Students Need More Sleep and Later School Start Times

By Thomas Franich, Christina Cardiel, Alia Carson, Marcos Martinez, Kian Mendoza, Sakari Morris, Christian Ramirez Acabal

In February, while going to the library to check out a manga comic one of my students recommended, I (Thomas Franich) grabbed a copy (as I always do) of *Scientific American* and read the February editorial "Let Teenagers Sleep." The editors advocated for delaying school start times based on overwhelming evidence that teens are biologically driven to go to bed later and wake up later than other age groups. The editors pushed for that change because later start times were linked to students' well-being and possible improvements in academic performance.

I was excited to hear others endorsing what I've argued for more than a decade, because, as a high school teacher, I know this is an issue that is important to my students. They are people who rarely (if ever) have input on the policies that shape their everyday lives.

This is why I gave the piece to the students in my physics and biology classes and asked them to read it and compare the claims to their experiences. We had a highly engaging conversation about the subject, and so I asked them to write letters to the editors at *SciAm* to share their thoughts.

What they wrote was illuminating.

I learned that my students are exhausted from the moment they get up in the morning, around 6:00 a.m. on average, until the moment they finally go to sleep, many at about 1:00 a.m. the next day. They told me that they can't fall asleep for hours, even though they feel foggy, leaden and tired. But, a few days later, we had a two-hour snow delay. My students got to sleep in. Later in class, they told me they felt refreshed, alert and aware, and they attributed the change to getting more sleep. I noticed changes too. They were more attentive, more engaged and more connected. Ancient as I am to them at 37, even I struggle to get up on time, so if every day was as on point as that one, I'm sure my students would integrate new knowledge far better than they do now.

We need data-driven school schedules, not just at Sandia High School, but throughout the Albuquerque public school district. And, luckily for our students, our district is listening. Sandia students, who have started school at 7:25 a.m. for many years will start next year at 8:40 a.m. The rest of the district's high schools will also start later, meaning the 23,000 high schoolers in our district will get the extra rest their bodies require.

In general, the students are energized by this announcement. They are highly motivated academically, and they think they'll earn better grades. They think they'll be more "chill" next year. Most importantly, they think they will learn better and retain more knowledge. Some are apprehensive about how their extracurriculars will be affected or how their employers will adjust their after-school schedules. A few students are concerned about how their families will adjust to the new schedule because they have siblings who start school at a different time. Yet, while there are nuances to settle, I agree with our superintendent who wrote in a letter to our community: "In a year or in three ... there won't be a push to go back to earlier start times for our ... students."

My students are hopeful for a better future, and I'm eager to see how the data from our schools will change in the next few years. Public policy should be informed with scientific data whenever possible and even though this is decades late, I am so happy that our students' dreams are finally becoming reality.

But don't take it from me. Here is what a few of my students have to say on sleep, school start times and student life:

Christina Cardiel, 9th Grade

Next year with the school time pushback, I truly believe students will be able to get enough sleep, which is one of the most important times for the body to grow and repair. Even with the new school start time, though, parents could let their kids take a 30-minute nap after school to reenergize and make up for the sleep they may have missed during the night. I'm glad that a decent time was set, so parents can still pick up and drop kids off while still being able to go to work. I think a school day of 8:40 a.m. to 3:40 p.m. is perfect to be able to function and learn throughout the school day and with an early enough ending time to do homework and rest—even with after-school activities.

Alia Carson, 9th Grade

If you're walking through the hallways of my high school on a typical day, you will almost certainly hear pained mutterings of "I'm tired," again and again. On an average day, many students have difficulty concentrating in early morning classes, and tend not to absorb the information we will later be tested on. When school start time was delayed an hour or two by weather a few months ago, my peers and I felt much more engaged and attentive. Although a relatively small shift in schedule, this hour could have far-reaching benefits. My own school district has just adopted a delayed start time for the new school year. Hopefully, this can serve as an example to other districts that this change is both possible and desirable. Perhaps being more stentorian about the issues discussed in "Let Teenagers Sleep" and explaining them unambiguously with data-supported urgency to politicians (who tend to bring up issues such as suicide, obesity and low test scores in campaign speeches) will drive policy supporting a healthier school schedule.

Marcos Martinez, 11th Grade

I believe the idea that giving teenagers more time to sleep, and the data supporting it, is an important step in bettering our education system. I hope that other school districts realize the significance of rest, so that my classmates and I and future generations may lead healthier and happier lives through our teen years. I find myself relieved that my future schedule will allow me to sleep in a little more, and I am not as worried about having to constantly learn while the only thing on my mind is sleep. I was happy to do Mr. Franich's assignment and write the letter to *Scientific American,* and I am proud to have a teacher who cares for me and my friends' mental health. Sometimes trying to make change can be daunting, especially when it comes to policies and procedures made by adults. By having an adult show me how I can make change gives me hope and a newfound passion to make my world better.

Kian Mendoza, 11th Grade

After so many years of research and evidence, it is clear we not only deserve, but need change. I have struggled with a severe lack of sleep throughout middle school and high school, and I've observed many of my peers, day after day, talk about how tired they are and how they only slept for so few hours. Not only has lack of sleep affected my mood and alertness at school, but I believe it has affected my academic performance. When I started having sleep problems in middle school, I started experiencing trouble with forgetfulness, depression, and my grades slowly began to decline. This has continued into high school, despite my, my doctor's and my therapist's best efforts. Most teens do not party every night or stay up late despite stereotypes. This is not about politics, it's about teenage health. Our district policy has changed our start time from 7:25 a.m. to around 8:40 a.m. for high school. For a long time, because of bus driver issues, high school had to start earlier, but why can we not have elementary school start times be earlier? Younger children's natural sleep cycle is earlier than most teens. Our school board members and I acknowledge that this change will affect many ways in which our school system runs, but it should overall benefit the mental and physical health for teens and teachers and push our students an extra step towards a more effective education. I'd like to see this change be national, so that all American youth can have this opportunity for a healthier life and future.

Sakari Morris, 11th Grade

When I get more sleep I have more energy and more than usual, I want to do my schoolwork. I also am in a way better mood and sleep better the next night, too. When we recently had a two-hour snow delay, there was a better atmosphere in classes and in the hallways, plus everyone was doing their work and paying more attention in class. I am looking forward to the next school year because I think it'll be a lot better for us to start at 8:40 a.m. I also think attendance is going to be better because people won't be as late or skip first period as much. I feel really proud about my advocacy for later start times because I feel like I'm doing something different for my junior year and that I'll be part of something that might make a difference. It's also cool because I feel like I can voice my opinion and it will be listened to and it won't be treated like I'm just another kid who doesn't know what they're talking about or that I'm just overreacting.

Christian Ramirez-Acabal, 11th Grade

Who knew sleep had such an importance on our mental health? I certainly did not, I just thought sleep was just a way to regain our energy. I was severely depressed from 7th grade to sophomore year. Some days I didn't want to even get up from bed. My mother would let me sleep an extra hour when it got so severe, I would have mental breakdowns. But that one hour of sleep helped way more than I can express in words; I felt happy, alive, willing to go to school, feeling like I could do anything, write a 10-page essay and breathe. Recently, the board of education passed a new thing where they moved the starting hour of school from 7:25 a.m to 8:40 a.m. A lot of my peers are angry about this, saying that their jobs are more important. Even my mom is upset about this because she is worried about getting to after school appointments. But I feel like this could benefit us greatly in our education and allow our mental health to recover. And if it doesn't, it was a great experiment to try.

This is an opinion and analysis article, and the views expressed by the author or authors are not necessarily those of Scientific American.

About the Authors

Thomas Franich is the biology and physics instructor at Sandia High School in Albuquerque, N.M.

Christina Cardiel is a high school student in New Mexico.

Alia Carson is a high school student in New Mexico.

Marcos Martinez is a high school student in New Mexico.

Kian Mendoza is a high school student in New Mexico.

Sakari Morris is a high school student in New Mexico.

Christian Ramirez Acabal is a high school student in New Mexico.

A Study Power Move: Sleep between Sessions

By Steven C. Pan

A good night's sleep can be transformative. Among its benefits are improved energy and mood, better immune system functioning and blood sugar regulation, and greater alertness and ability to concentrate. Given all of these benefits, the fact that a third of the human lifespan is spent sleeping makes evolutionary sense. However, sleep appears to have another important function: helping us learn. Across a plethora of memory tasks–involving word lists, maze locations, auditory tones, and more–going to sleep after training yields better performance than remaining awake. This has prompted many sleep researchers to reach a provocative conclusion: beyond merely supporting learning, sleep is vital, and perhaps even directly responsible, for learning itself.

Recent discoveries from neuroscience provide insights into that possibility. Sleep appears to be important for long-term potentiation, a strengthening of signals between neurons that is widely regarded as a mechanism of learning and memory. Certain memories acquired during the day appear to be reactivated and "replayed" in the brain during sleep, which may help make them longer lasting. In some instances the amount of improvement that occurs on memory tasks positively correlates with the length of time spent in certain stages of sleep. These and other findings are generating great excitement among sleep researchers, as well as prompting heated debates about the degree to which sleep may or may not be involved in learning.

To date, most sleep and learning research has focused on recall, which is the capacity to remember information. However, new research by Stéphanie Mazza and colleagues at the University of Lyon, recently published in the journal *Psychological Science*, suggests another potential benefit of sleep: improved relearning. Relearning refers to the process of re-acquiring forgotten information.

Because we cannot possibly remember all of the information that we encounter, it is often necessary to go back and learn that information again. That's when relearning occurs–such as when preparing for an exam, taking a refresher course, or simply revisiting a topic after an extended period of time. According to this new research, sleep supercharges relearning: it can enable us to relearn twice as quickly and up to three times more effectively.

In the study, 40 French-speaking adult participants learned a list of 16 Swahili-French word pairs (for example, *nyanya-tomate*), relearned those word pairs after a period of 12 hours that did or did not include sleep, and were tested on their memory for the word pairs after one week and after six months. Each participant completed the study as part of a randomly-assigned "wake" or "sleep" group. In the wake group, the initial learning session occurred at 9 a.m. and the relearning session occurred at 9 p.m. on the same day; in the sleep group, the initial learning session occurred at 9 p.m. and the relearning session occurred at 9 a.m. on the next day, after a night of sleep. During the learning session, participants first studied all of the word pairs. Next, the Swahili word from each pair was shown by itself while participants attempted to recall its French equivalent. After each attempt, the correct answer was displayed, allowing participants to study further. The learning session concluded after each pair was successfully recalled once. At the relearning session, participants again attempted to recall the French word of each pair, and then continued cycling through the entire list until they could recall all pairs perfectly. Thus, by the time the relearning session was complete, any word pair that had been forgotten in the intervening 12 hours had been practiced until it could be correctly recalled once more.

During the initial learning session there were no observed performance differences between the sleep and wake groups, which suggests that a similar degree of learning occurred in both groups. However, substantial differences emerged at the relearning session: participants in the sleep group recalled an average of 10 word pairs on their first attempt, whereas those in the wake group could only muster up about 7; in addition, those that had slept took only about three cycles through the list to finish relearning, whereas those that had not slept needed twice that amount. Even the lowest-performing members of the sleep group–those that had forgotten more word pairs than other members and had the most to relearn–still took fewer cycles to relearn than the best performing (and least forgetful) members of the wake group.

Along with that impressive result, a benefit of having slept between the learning and relearning sessions was also evident over the long term. After one week, participants in the sleep group correctly recalled more word pairs than those in the wake group (an average of 15 vs. 11), and after six months, their recall performance, while attenuated due to some forgetting, was three times better than that of the wake group (an average of 9 vs. 3). Thus, despite the fact that both groups had practiced to a level of perfect recall during the relearning session, only those that had slept prior to relearning accrued any long-term memory benefits.

The fact that participants in the sleep group engaged in relearning was also critical for their improved long-term performance. In a follow-up to the main experiment, a control group of 20 additional participants followed nearly all of the same procedures as the sleep group, including training at the same times of day and sleeping after initial learning. However, in place of the relearning session, these participants only briefly practiced the list of word pairs one time. Without relearning, their subsequent recall performance was no better than the wake group. Thus, it was the combination of sleeping after initial learning, plus engaging in relearning itself, that generated long-term memory benefits.

Mazza and her colleagues interpreted their results as evidence that sleep "transforms" memories, making the effects of initial learning stronger and empowering subsequent relearning. By this interpretation, sleep not only prevents memories from being forgotten, it also makes it easier to restore memories during relearning. In support of this conclusion, the authors noted that total sleep time was positively correlated with recall and relearning performance (the longer participants slept, the better they tended to do on both). This pattern, similar to that observed in prior research, is consistent with two possibilities: either a sleep-specific mechanism boosts learning, which sets the stage for improved relearning, or sleep allows regularly occurring learning processes to occur without interference, which also boosts memory and expedites relearning. Both possibilities remain viable in this and related research, thus inviting future studies to tease the two apart.

Moreover, while this latest study reveals an important and largely unknown potential benefit of sleep, it remains to be determined whether the same benefits will manifest in other circumstances (such as when relearning non-verbal materials). It is also important to note that Mazza and her colleagues had their sleep and wake groups train at different times of day. In a recent meta-analysis of sleep and motor learning research, Tim Rickard and I found that such "varied time" experimental designs often yield better performance in sleep vs. wake groups regardless of whether there is a sleep-specific learning benefit. This may be due to the fact that humans vary in fatigue levels, alertness, and other characteristics according to the time of day and their natural urge to sleep—differences that may not necessarily manifest in the sleepiness surveys that participants of sleep studies are often asked to take (as occurred in the present study). Further research in which the sleep and wake groups are not subjected to such widely divergent schedules is needed to address this issue.

At present, these new results suggest that sleeping between study sessions can be a particularly potent learning strategy, and especially when learners take the time to repeatedly revisit and relearn in order to achieve a high level of performance. For instance, if a set of materials needs to be completely mastered, such as an official manual in preparation for a licensure exam, then those materials should be repeatedly studied across multiple study sessions and in between periods of sleep. Scheduling those study sessions to occur shortly before going to bed is also likely to improve future remembering and relearning. In a world where an uninterrupted night of sleep seems to be increasingly rare—and many of us try expensive and unproven ways to improve learning and memory—these may be some of the most compelling reasons yet to embrace the Land of Nod.

About the Authors

Steven C. Pan is a chancellor's postdoctoral fellow and learning scientist at the University of California, Los Angeles. His research involves using the tools of cognitive psychology and neuroscience to enhance human learning, memory and performance.

Sound Awake: "Noisy" Neurons May Repeatedly Disrupt Your Sleep

By Dana G. Smith

You don't remember it, but you woke up at least 100 times last night. These spontaneous arousals, lasting less than 15 seconds each, occur roughly every five minutes and don't seem to affect how well-rested you feel. They are unrelated to waking up from a bad dream or your partner tossing and turning. Instead, they seem to be linked to some internal biological mechanism.

Frequently waking up throughout the night may have protected early humans from predators by increasing their awareness of their surroundings during sleep. "The likelihood someone would notice an animal is higher [if they] wake up more often," says Ronny Bartsch, a physicist at Bar-Ilan University in Israel. "When you wake up, you're more prone to hear things. In deep sleep, you're completely isolated."

Sleep scientists, however, have been stumped as to what triggers these nocturnal disruptions. In a new *Science Advances* paper Bartsch proposes an innovative hypothesis that spontaneous arousals are due to random electrical activity in a specific set of neurons in the brain—aptly named the wake-promoting neurons.

Even when you are asleep your brain cells continuously buzz with a low level of electrical activity akin to white noise on the radio. Occasionally, this electrical clamor reaches a threshold that triggers the firing of neurons. The new paper suggests that when random firing occurs in the wake-promoting neurons, a person briefly jerks awake. But this is countered by a suite of sleep-promoting neurons that helps one quickly fall back to sleep.

Low-level electrical activity in neurons increases in colder temperatures whereas warmer temperatures flatten it. As a result, there should be fewer spontaneous arousals in hot weather. To test this theory, the researchers created computer models that mapped how neuronal noise should act at different temperatures and how the varying electrical activity could affect spontaneous arousals. They also measured sleep in zebra fish, which have similar day/night cycles to humans but are ectothermic, meaning their body temperature is controlled by the environment rather than by internal processes.

The researchers compared the fish's sleep rates at four different water temperatures: 77, 82 (ideal for zebra fish), 84 and 93 degrees Fahrenheit. Across the board, the colder the water the more often the zebra fish woke up and the longer they stayed awake. The data from the zebra fish and the models of temperature, neuronal noise and arousal matched perfectly. "I think their theory is a perfectly good one and may even be correct," says Clifford Saper, a neuroscientist at Harvard Medical School's Division of Sleep Medicine and head of neurology at Beth Israel Deaconess Medical Center who was not involved with the study. "But the experiment they did doesn't test that hypothesis."

The zebra fish experiment shows the fish wake up more frequently and stay awake for longer in colder temperatures but reveals nothing about these animals' neuronal noise–or humans', for that matter. Bartsch says that, so far, no studies have figured out how to measure neuronal noise in a sleeping animal.

The idea that warm temperatures cause fewer nocturnal disruptions also seemingly flies in the face of conventional wisdom that a colder bedroom leads to better sleep. But waking up because you are hot and uncomfortable is different from these brief spontaneous arousals. In fact, our bodies are pretty good at regulating their core brain and body temperatures, so the difference of a few degrees outside would not alter neuronal activity. In contrast, zebra fish's temperature varies quite a bit. Saper says because of this zebra fish "are probably the last animal that I would use to try to make this point."

Bartsch emphasizes the study is not trying to make a claim about thermoregulation in adults but he says it may have implications for newborn babies. "Because very young infants are more ectothermic than endothermic, their arousability could scale similarly to fish for different ambient temperatures."

Infants are not as good at regulating their own temperature and so are more vulnerable to changes in the environment. (This is why premature babies have to be kept in incubators.) Consequently, the researchers think newborns may be more susceptible to heat-related fluctuations in neuronal noise.

The theory may have important implications for infant sleep. Although they may be disruptive to parents, spontaneous arousals could help save a baby's life. Sudden infant death syndrome (SIDS) has been a leading cause of mortality in children between one month and one year of age and yet largely remains a mystery. One idea is that SIDS is caused by a stoppage in breathing, often through accidental suffocation. Waking up during the night can prompt babies to shift or cry out, helping to ensure that they do not have anything obstructing their airways and are still breathing. "We came up again with a theory that the babies with SIDS have low neuronal noise and therefore they have lower arousals," says Hila Dvir, a physicist at Bar-Ilan University who co-authored the paper. "Because they have low arousals, they are less protected from any hypoxic event–a shortage of oxygen."

Not everyone is convinced, though. "Over the years, people have come up with ideas to explain SIDS, like a single explanation for it, and they just keep hitting dead ends with it because it's probably a complex, heterogeneous situation," says Rafael Pelayo, a clinical professor at the Stanford Center for Sleep Sciences and Medicine. "It is a cool idea that this neuronal noise is explaining the arousals. I just think they jumped a little bit when they got into SIDS. It has to be more complicated than that."

About the Author

Dana Smith is a freelance science writer specializing in brains and bodies. She has written for Scientific American, the Atlantic, the Guardian, *NPR, Discover, and Fast Company, among other outlets. In a previous life, she earned a Ph.D. in experimental psychology from the University of Cambridge.*

Should You Give Your Kid Melatonin?

By Elana Spivack

Earlier this year in Indiana, day care director Tonya Rachelle Voris was fired and arrested for feeding melatonin supplements to young children in the New Life Church's childcare program without parental consent. Voris now faces 11 counts of neglect of dependents and six counts of reckless supervision by a child care provider.

The use of melatonin, a common sleep aid, has increased significantly in the last two decades. More than five times as many adults in the U.S. took melatonin in 2018 than in 2000. Kids are also exposed to or given melatonin: a 2020 study found that 1.3 percent of children and adolescents take the supplement. But although melatonin is an accessible and easy method to help with sleep, it isn't always the most effective remedy, especially for children.

In the U.S., melatonin is categorized as a dietary supplement and thus doesn't require a prescription, whereas in the European Union, the U.K., Australia, Japan and Canada, it is available by prescription only. For kids, there's no universally agreed-upon dosage or age at which they can start taking the supplement. With its availability over the counter in drug stores and supermarkets and inclusion in a variety of flavored gummies, melatonin is easier to use–and misuse–than ever before. Scientific American reached out to experts to discuss the way melatonin works in the body, its proper dosages and its appropriate use for children.

What Is Melatonin?

Melatonin is a hormone produced in the brain in response to darkness that signals that it's time to sleep. This neurochemical comes from the pineal gland, a petite structure located in the center of the brain that philosopher René Descartes once described as the "seat of the soul" and rational thought. The gland produces about 0.3 milligram of melatonin per day and has no other known functions.

This bedtime hormone isn't a soporific (sleep inducer) but a part of the circadian rhythm, a person's internal biological clock that's tuned in to the sleep-wake cycle. Once produced, the hormone enters the bloodstream and fluid surrounding the brain and spinal cord. The body produces 10 times more melatonin at night than during the day, so when melatonin receptors detect this peak, they register that it's nighttime.

What Are Melatonin Supplements Used For?

Consumers in the U.S. primarily use over-the-counter melatonin supplements as a sleep aid. These supplements, including sugary gummies, typically offer much higher doses than what the pineal gland produces to help shift circadian rhythm to night mode. Melatonin dosages will affect people differently, and more research is needed to understand the most effective amounts to assist sleep. Some research does show that supplemental melatonin works best when taken two hours before bedtime, as opposed to right before sleeping.

One serving may be listed as containing as little as 0.1 mg and up to 20 mg of the hormone, though 3, 5 and 10 mg per serving are the most common amounts. Getting that intended dose may be challenging, however: a study recently published in *JAMA* found that in 25 brands of over-the-counter melatonin gummies, the actual dosage significantly differed from what was advertised. One type of gummy was found to contain almost 3.5 times the labeled quantity of melatonin.

Some studies suggest that melatonin supplements may be the most help to adults suffering from jet lag or a wrecked sleep schedule. They are not the best remedy for someone with insomnia or a night shift worker. "There [are] just some preliminary studies that suggest [melatonin] might be helpful in some certain specific situations, but even in those situations, the data is not definitive," says Pieter Cohen, lead author of the *JAMA* study and a general internist at the Cambridge Health Alliance in Massachusetts.

Cohen says that the best indicator of whether a labeled dosage is trustworthy is third-party certification from nonprofit organizations that test final batches of a company's product for quality, such as the U.S. Pharmacopeia and NSF (formerly the National Sanitation Foundation).

Can Children Take Melatonin Supplements?

From tots to teens, kids don't always have the most regular sleep habits. Melatonin gummies may seem like an easy way to regulate bedtime for both stress-addled high schoolers and wired toddlers, but experts say they shouldn't be a constant strategy to fall back on.

Melatonin is an active hormone that should be taken more seriously than "a glass of warm milk or singing a lullaby to help someone go to sleep," Cohen says. While our bodies produce the hormone naturally, careful consideration must go into the decision to take it as a supplement, especially for kids.

"There are other, more effective ways of getting kids to sleep," says Ken Sassower, a pediatric neurologist at Massachusetts General Hospital. And as with other supplements, more doesn't necessarily mean better. If a low-dose gummy doesn't do the trick, popping another one probably won't either.

Melatonin is, however, used to help safely regulate sleep in children with attention deficit hyperactivity disorder (ADHD) and autism spectrum disorder (ASD), both of which are associated with circadian rhythm disturbances. "It's been accepted in the management of autism sleep issues," Sassower says. While recent studies find that melatonin may be a safe and effective treatment for these conditions, researchers such as Sassower say this use requires more study.

In kids without conditions such as ADHD and ASD, there's not much evidence-based guidance regarding the best age for use. The particular brand of melatonin gummies that Tonya Rachelle Voris gave at the day care in Indiana was recommended for kids age four and older. The children she gave the supplements to were one to four years old.

According to Phyllis Zee, a sleep neurologist at Northwestern University's Feinberg School of Medicine, while "there is no data on which age group should or should not take melatonin," caretakers should make a joint informed decision with a health care professional on a case-by-case basis. She explains that doses below 1 mg appear to be safe for kids.

Is It Possible To Overdose on Melatonin?

The concept of a melatonin "overdose" remains controversial depending on how one defines an overdose. Excessive amounts of the hormone may result in many different side effects, including nausea, vomiting, headaches, mild hypothermia, fatigue, confusion, nightmares and low blood pressure. It's extremely unlikely that taking high doses of melatonin will cause death. In one case, a 42-year-old woman survived after ingesting 120 mg and receiving emergency treatment, including a stomach pump. But for children, too much melatonin can cause serious, even life-threatening, poisoning. Between 2012 and 2021, two children died out of more than 27,700 who were treated for pediatric melatonin ingestion at health care facilities.

Children will always have a lower threshold for overdose than adults because of their size. And they may be enticed by the fruity gummies, which could increase their risk of ingesting higher amounts of melatonin if they are not carefully monitored. Cohen says that if a child has intentionally or unintentionally eaten more than twice the recommended serving size of melatonin for their age and size, that merits a call to a pediatrician.

What's more, many melatonin supplements also contain cannabidiol (CBD), an ingredient that is active in cannabis and usually derived from hemp. Cohen says children should never take CBD. Combination melatonin-CBD supplements may also have much higher levels of the cannabis compound than advertised, as shown in Cohen and his colleagues' recent *JAMA* study. In one of the study samples, a combination gummy supplement contained no melatonin, only CBD.

Our bodies may naturally produce melatonin, but taking additional amounts of the hormone should be done with caution. Melatonin supplements have been characterized as one of the "least toxic" medications. Adults and caregivers still should not use them recklessly and should be aware of the side effects.

"As a parent, I can tell you I wouldn't give melatonin to my kids unless my pediatrician recommended that I do that," Cohen says.

About the Author

Elana Spivack is a science journalist in New York City who covers reproductive health, food science, and more. Follow her on Twitter @ElanaSpivack.

Section 5: Snoozing Animals

Memory Wins When Dogs Sleep

By Julie Hecht

A Harvard Medical School professor recently rocked the Internet: "Since dogs are generally extremely attached to their human owners, it's likely your dog is dreaming of your face, your smell and of pleasing or annoying you," psychologist Deirdre Barrett told *People* magazine.

And then hearts everywhere exploded.

Barrett's sleep research focuses on humans, while an interest in evolutionary psychology helps her consider the sleep of non-human mammals. Both have similar sleep cycles, she notes, which could suggest parallels in sleep quality or experience.

But an open access study in *Scientific Reports* out yesterday moves away from extrapolation and toward hard data. Researchers in Hungary have devised a way to non-invasively peer into the sleeping dog's brain to explore the content and function of their sleep.

Sleep in dogs is good for a number of things, including, but not limited to cuteness, cuteness, and more cuteness. But you've also probably heard that sleep is good for memory. Before a big test we're often told, "Get a good night's rest," which is actually shorthand for–give memory consolidation a chance. "Memory consolidation" is the process where your brain pulls together pieces of information and packages them into memories that can be used in the future.

Memory is also important for dogs. Working dogs need to learn–and retain–a wide variety of job-specific skills, and companion dogs often learn basic skills to successfully live alongside humans. When a dog learns something new, can sleep help the dog perform those skills better? Should training sessions incorporate naptime?

Anna Kis of the Hungarian Academy of Sciences and colleagues–including members of the well-known Family Dog Project–set out to explore the relationship between sleep and memory in companion dogs. Their study involved two experiments: the first gave dogs a learning task and then peered into their sleep via non-invasive electroencephalogram (EEG)–a test that detects brain electrical activity using small electrodes attached to the scalp. The second experiment explored whether different type of post-learning activities (such as sleep) affect memory consolidation, both in the short- and long-term. All experiments were performed with consenting companion dogs and their helpful owners.

First up, the sleep study, also known as polysomnography if you want to be fancy about it. Fifteen companion dogs participated in both a learning and a non-learning condition. The experimenters taught the dogs the commands for "sit" and "lie down" in a foreign language (English). As you'd expect, no learning took place in the non-learning condition–dogs simply practiced the "sit" and "lie down" commands that they already knew in Hungarian. Nothing new. Old hat. (Most dogs don't wear hats. Old collar?)

For the critical phase of the experiment, dogs went to sleep (gosh I love science). Dog snoozing-related brain activity was then monitored over the next three hours. Afterwards, dogs in the learning condition were retested on "sit" and "lie down" in English to determine whether sleep helped the dogs process what they had learned.

Not only did the sleep affect dogs' learning, the learning affected dogs' sleep. Dogs did better responding to "sit" and "lie down" in English after taking a snooze. But even before the dogs in the learning condition were retested, two notable wave patterns stood out in the EEG spectrum in the non-REM phase (the dreamless part of sleep). There was an increase of delta power, similar to what is found in humans, and a decrease in alpha activity, which could suggest "an increase in sleep depth after learning."

These two findings are related. Dogs learned a task, which alters their brain activity during sleep, then they performed better on the task. "This suggests that the newly acquired information is re-processed and consolidated during sleep," Kis explained over email. More specifically, the correlation between the post-sleep improvement in performance and certain EEG patterns "is the strongest indicator that the changes in sleep EEG we see after learning are functionally related to memory consolidation," added Kis.

Neat. Taking a snooze can improve subsequent performance (at least for this type of command learning task). But how do we make things stick? Is sleep more or less effective than other strategies for retaining information? A second behavioral experiment investigated the effect of different post-learning activities (including sleep) on subsequent memory.

Fifty-three new companion dogs learned "sit" and "lie down" to new words (again, English). Dogs were then put in one of four different post-learning groups, spending the next hour either sleeping, walking, learning more (learning new behaviors via the luring training method), or eating from and playing with a Kong dog toy. When the hour was up, dogs were retested on the English commands they'd just learned.*

The type of post-learning activity seemed to affect dog performance in the short term, but not exactly as the researchers had expected. In the short term, both sleeping and walking improved subsequent performance, while more learning and Kong play did not.

On the other hand, when dogs came back a week later, presumably after many sleeps, dogs in the sleep, walk, and Kong play conditions showed marked improvement with the English commands. Dogs who had done more learning did not improve.

Dog lovers often think about learning and obedience in terms of dogs doing it "right" or "wrong." Factors surrounding learning, this study reminds, can affect memory consolidation and later performance.

Kis recommends: "Learning a new command should be followed by an activity that does not interfere with this new memory trace (e.g. sleeping, walking, playing–but not learning other things) in order to achieve the highest subsequent performance in the long run."

At the same time, Kis noted that dogs in the sleeping condition might have performed even better if the nap extended beyond an hour (possibly for memory consolidation to fully take place), or if, after waking up, the dogs had a few more minutes to shake off their sleepiness before performing the tasks again. Human-sleep scientists refer to this latter phenomenon of decreased cognitive performance in the few minutes after waking up as "sleep inertia." Don't pretend you've never woken up, walked to the bathroom, and tried to brush your teeth with your comb. Since no sleep inertia interval has been established for dogs, Kis says, they can't rule out the possibility that the dogs were still sleep zombies when they were retested.

Non-invasive studies of dogs and sleep are new. We haven't yet studied whether your dog is dreaming of your face or your glorious smell, but if you care about learning in dogs, this study suggests you give sleep a chance.

* *Maybe you're wondering why there wasn't a condition after learning where dogs simply rested–rather than slept–and then had their memory tested. This "resting" awake condition is typically found in human memory consolidation studies because it's the closest match to the 'sleep' condition. But this condition was not included for dogs, the researchers explain, "as preventing dogs from falling asleep while requested to stay in a laying position for one hour would presumably induce stress in the animals. Stress is known to have an impact on memory, and also raises animal welfare issues, thus we decided to avoid such a condition."*

The views expressed are those of the author(s) and are not necessarily those of Scientific American.

Referenced

Kis et al. (2017). "The interrelated effect of sleep and learning in dogs (*Canis familiaris*); an EEG and behavioural study." *Scientific Reports* 7, 41873.

Kis et al. (2014). "Development of a non-invasive polysomnography technique for dogs (*Canis familiaris*)." *Physiology & Behavior*, 130, 149–156.

Affenzeller et al. (2017). "Playful activity post-learning improves training performance in Labrador Retriever dogs (*Canis lupus familiaris*)." *Physiology & Behavior*, 168, 62–73.

About the Author

Julie Hecht is a Ph.D. student studying dog behavior and writes the Dog Spies blog *at ScientificAmerican.com.*

Meet the Animals That Literally Sleep with One Eye Open

By Gian Gastone Mascetti

One of the most striking features of living organisms, both animals and plants, is the way their physiology and behavior have adapted to follow the fluctuations of daily light and nocturnal darkness. A clock in the brain synchronized to environmental cues generates biological changes that vary over a 24-hour cycle—circadian rhythms (from the Latin words *circa* and *diem*, meaning "about" and "a day," respectively). In this way, the earth's rotation is reproduced in the dynamics of our neuronal circuits.

The sleep-wakefulness cycle is a typical circadian rhythm. Wakefulness is characterized by sensory activity and movement; during sleep the senses lose touch with their surroundings, and movements subside. This periodic loss of consciousness appears on electroencephalogram (EEG) recordings as a clear signature: deep sleep consists of slow oscillations of high amplitude. Wakefulness, in contrast, is made up of fast, low-amplitude oscillations. Much about sleep remains a mystery, however. Why would an animal shut down basic sensory and motor activity for hours on end, leaving itself a target for predators? This question becomes more acute in aquatic mammals, which need to regulate breathing and body temperature while they sleep.

Remarkably, some animals have solved this problem by developing the ability to sleep with one half their brain while remaining vigilant with the other—a behavior known as unihemispheric slow-wave sleep (USWS). Still others engage in USWS under some circumstances but put both hemispheres to bed when necessary. Marine mammals, bird species and possibly reptiles enter a half-on/half-off state, sometimes keeping one eye open during these intervals. Recently researchers have even discovered a vestigial form of unihemispheric sleep in humans.

Half-slumber provides a fascinating vista into the science of sleep. While studies are carried out on the dormant half, the opposite side can serve as the requisite control for experiments. The ability to thrive with a relative lack of sleep, as dolphins and some birds do, may provide ideas for treating human sleep disorders, which often affect one brain hemisphere more than the other.

Asleep But Not Really

The study of unihemispheric sleep started in 1964, when controversial researcher John C. Lilly suggested that dolphins could sleep using one side of the brain after observing that the animals keep only one eye closed during their daily rest. Lilly assumed that when asleep, dolphins could still watch and listen to their surroundings. It would take later experiments to determine what was happening in cetacean brains.

Cetaceans–whales, dolphins and porpoises–are still the subjects of studies on unihemispheric sleep. The animals preserve two physiological features from their ancestors' life on land: lungs for breathing air and mechanisms for maintaining nearly constant body temperature in water (thermoregulation). Sleeping with half a brain, it seems, has allowed them to retain those features in an aquatic environment.

More recently, Lev Mukhametov of the A. N. Severtsov Institute of Ecology and Evolution at the Russian Academy of Sciences and his colleagues looked more deeply than Lilly did into what was happening in the cetacean brain. Mukhametov and his colleagues studied sleep extensively in bottlenose dolphins. In EEG recordings, the researchers consistently found that one hemisphere of the animals' brain was in a state of slow-wave sleep, while the other was awake. They rarely observed sleep in both hemispheres (which is called bihemispheric slow-wave sleep, or BSWS), and they recorded no unequivocal signs of the rapid eye movement (REM) sleep associated with dreaming.

During USWS the awake hemisphere of a dolphin's brain controls swimming and surfacing to breathe. As Lilly surmised from cursory observation, the animal's one open eye, linked to the contralateral awake hemisphere of the brain, allows a dolphin to monitor for predators and swim in unison with its companions while the other half of the brain rests. In 1999 P. Dawn Goley of the department of biological sciences at Humboldt State University observed—as did Guido Gnone of the Aquarium of Genoa in Italy and his colleagues in 2001—that when dolphins swam in groups, the open eye of a pod member maintained visual contact with others. If a partner shifted to the opposite side, the eye pattern reversed.

Dolphins also confront cold water temperatures that expose them to high heat loss. Keeping one hemisphere of the brain awake during rest allows the animals to stay warm by frequently moving their flippers and tail to swim and hover near the surface while they sleep—observations reported by Praneshri Pillay and Paul R. Manger, both then at the University of the Witwatersrand, Johannesburg.

We know that in cetaceans and other animals, the overall sleep-wake cycle is governed by interactions among multiple brain structures, including the brain stem, the hypothalamus and the basal forebrain. Precisely what regulates unihemispheric sleep remains a mystery, although we have clues. In 2012 David J. Kedziora and his colleagues at the University of Sydney worked out a mathematical model of USWS intended to represent dolphin sleep habits. In it, substructures within the hypothalamus in each hemisphere—the ventrolateral preoptic nuclei—exchange messages to regulate when sleep occurs in each hemisphere. It appears that inhibitory signals transmitted between the two hemispheres could allow one side to go to sleep while the other stays awake. Deep-brain structures, such as the posterior commissures in the brain stem, would also be involved. (The posterior commissures are extremely large in dolphins, giving rise to questions about their role in managing sleep.) The University of Sydney model gives neuroscientists a way to explore the mechanisms of how the brain hands off the delicate task of allocating sleep to one hemisphere or another.

When immersed in water, northern fur seals sleep with one brain hemisphere while adopting a body posture that allows them to breathe and stay alert for approaching predators.

Environmental cues also seem to play a role. Because the sleep-promoting neurons in the hypothalamus are thermosensitive, a rise or fall in brain temperature causes a corresponding fluctuation in the firing rate of these neurons. Indeed, in 1982 Mukhametov and his colleagues found that during USWS dolphins' brain temperatures decreased in the sleeping hemisphere and remained constant in the awake one.

A Singular Adaptation

Cetaceans evolved from a common terrestrial ancestor with hippopotamuses and other hoofed mammals. The move from a terrestrial to an aquatic environment was gradual and may have included a semiaquatic transition that entailed significant physiological and behavioral adjustments. Consequently, cetaceans' sleep behavior represents a singular example of adaptation to a new environment that demonstrates a trade-off between the need for sleep and survival.

Other animals make similar compromises. Seals, for example, have adopted various evolutionary solutions to the closely related problem of breathing and sleeping in water and on land. Some families of seals eschew USWS altogether. It has not turned up in earless, or "true," seals (the family *Phocidae*), including harp and elephant seals.

Northern fur seals (the family *Otariidae*), however, demonstrate a different story. In 2017 Oleg I. Lyamin of the A. N. Severtsov Institute of Ecology and Evolution reported that unlike dolphins, which appear to rarely experience BSWS and perhaps never enter REM sleep at all, northern fur seals undergo multiple sleep types, including BSWS, REM and USWS, in both their aquatic and terrestrial lairs. On land, BSWS predominates. In water, the amount of time spent in USWS increases, compared with that on land. REM sleep in water diminishes or even disappears.

When immersed in water and experiencing USWS, fur seals adopt a body posture that allows them to sleep, breathe and track approaching predators: they lie on one side with one flipper in the water and paddle continuously with it while keeping their other three flippers in the air to reduce heat loss. Their nostrils, meanwhile, remain out of the water so the seals can breathe. The brain hemisphere on the opposite side from the moving flipper (and the one open eye) is awake, letting the animals issue motor commands for paddling and retaining postural stability. On land, USWS allows fur seals to watch for predators and coordinate activity with companions, but it does not help with breath control, body temperature or coordination of movement.

Some birds, too, engage in unihemispheric sleep as they balance the need for rest and defensive alertness. (At times, USWS is combined with BSWS and REM.) In 1996 Jadwiga Szymczak, then at Nicolaus Copernicus University in Poland, recorded the presence of slow-wave EEGs in one hemisphere of the European black bird. And in 2001 Niels C. Rattenborg, then at the department of life sciences at Indiana State University, and his colleagues did the same in pigeons. Similarly, in 1999 Rattenborg had found that mallard ducks sleep with only half a brain to watch for threats. Ducks that kept one eye open while stationed at the outside edges of a group showed 150 percent higher levels of USWS than birds located toward the center. The open eyes of the "sentinel" ducks were directed away from the group. Mark A. Elgar, now at the University of Melbourne in Australia, reported in a 1989 study that vigilance decreases when the group grows larger and when an animal moves toward the center of the group.

Migrating birds during nonstop, long-distance flights also rely on differing sleep strategies. In 2016 Rattenborg, now at the Max Planck Institute for Ornithology in Seewiesen, Germany, and his team studied USWS and BSWS in great frigate birds (*Fregata minor*) during their 10-day sojourns. In a single USWS episode, one hemisphere showed a waking EEG pattern contralateral to the direction of a flight turn, indicating that the open eye on the opposite side was watching where the flock was headed. Also, Thomas Fuchs, then at Bowling Green State University, discovered in 2006 that Swainson's thrush compensates for the loss of sleep during night flight by increasing total sleep time, taking daytime micro naps and, when perching, closing one eye.

What About Us?

Humans do not engage in classic USWS, but they occasionally experience something reminiscent of it. Masako Tamaki and her group at Brown University made EEG recordings when people spent the night in an unfamiliar environment. In a 2016 publication by Tamaki, the EEGs showed slow waves indicative of deep sleep in the right hemisphere and shallow slow-wave activity in the left hemisphere, a sign of more alertness. The left hemisphere, moreover, was more easily aroused than its opposite half. This asymmetry, referred to as the first-night effect, disappears by the second night but seems to preserve vigilance in an unfamiliar place. It is reminiscent of mothers who retain a low awakening threshold to cries or other noises they identify as coming from their babies.

We may feel the sleep debt after the first night away from home. But other animals that sleep with one side of the brain all the time seem to be well adapted to their routines. Those that immerse themselves in USWS spend less time sleeping, compared with those that engage in BSWS or REM sleep.

Even so, their ability to swim, fly, eat or socialize with companions remains undiminished. Dolphins spend almost two-thirds of their day awake and the rest of the time in USWS, trading off sleep time between the two hemispheres. Brain and body recovery, however, does not appear to be affected, despite the absence of REM sleep.

In 1997 Mukhametov and his colleagues reported that dolphins in sleep studies always appeared to be in good health. In captivity, where scientists could observe the animals closely, dolphins learned and memorized complex tasks. Frigate birds sharply reduced total sleep while flying but maintained a high level of attention and efficient flight performance during their extended journeys.

Some animals seem to cope by sharing the half-sleep burden. Mallard ducks that act as flock sentinels by keeping one eye open lose sleep but without impaired behavior. The birds later pass off their lookout roles to a companion on another day. Unihemispheric sleep continues to fascinate the research community because it illustrates the diverse evolutionary strategies that have emerged to allow animals to get their rest every day.

The interest generated in USWS from field experiments has even made it into a laboratory tool for exploring the role of sleep in helping to shape development of the brain just after birth. In 1999 my group in the department of general psychology at the University of Padova in Italy found that just-born chickens (*Gallus gallus*) experienced significantly more left-hemisphere sleep during the first week after hatching. The chicks favored that hemisphere in those early days to learn about stimuli—patterns and colors—that must be processed for the first time by their new brain: sleep appeared to play a role in organizing what they had just learned.

Right-hemisphere sleep increased in the chicks as activity such as spatial analysis and the processing of new events prevailed on that side in the second week. When we trained chicks in a color-discrimination task, they subsequently registered more left USWS (with their right eye closed and their left hemisphere asleep) because that hemisphere was dominant in learning about colors. Chicks used the left eye for a spatial-learning task that required them to select one of four containers in a particular corner of their enclosure. They had to pick the container with a hole on top that contained a food treat. When they were done, chicks showed more right USWS (with their left eye closed and their right hemisphere sleeping) to rest the side of the brain that specializes in this type of task.

The most active hemisphere—whether engaged in USWS or BSWS—spent relatively more time sleeping to allow for recovery. Meanwhile the open eye on the side of the nondominant hemisphere took over to watch for predators and to stay apprised of the environment. In fact, moving a dark object over the cage during USWS caused chicks to wake immediately, startle and emit distress calls. Vigilance remained intact, but it did not detract from sleep as a time to sort out the intense sensory experiences of the birds' first days in a new world.

Ultimately studying animals that sleep with half a brain could aid us in understanding the continuing biological enigma of sleep—and perhaps even sleep problems in humans. Apnea and other disorders sometimes have effects more in one hemisphere than the other. This work may help answer how a species balances the benefits of rest with a need to protect itself against a hungry predator. Sleeping with one side of the brain is a brilliant answer to this dilemma, enabling an animal to experience conscious and unconscious states all at once. Research on unihemispheric sleep resonates through the millennia with a frequently quoted passage from Heraclitus' *Fragments*:"Even a soul submerged in sleep is hard at work and helps make something of the world."

Referenced

Cetacean Sleep: An Unusual Form of Mammalian Sleep. Oleg I. Lyamin et al. in *Neuroscience & Biobehavioral Reviews*, Vol. 32, No. 8, pages 1451–1484; October 2008.

Physiologically Based Quantitative Modeling of Unihemispheric Sleep. D. J. Kedziora et al. in *Journal of Theoretical Biology*, Vol. 314, pages 109–119; December 7, 2012.

Unihemispheric Sleep and Asymmetrical Sleep: Behavioral, Neurophysiological, and Functional Perspectives. Gian Gastone Mascetti in *Nature and Science of Sleep*, Vol. 8, pages 221–238; 2016.

Evidence That Birds Sleep in Mid-Flight. Niels C. Rattenborg et al. in *Nature Communications*, Vol. 7, Article No. 12468; August 3, 2016.

Sleep in the Northern Fur Seal. Oleg I. Lyamin et al. in *Current Opinion in Neurobiology*, Vol. 44, pages 144–151; June 2017.

About the Author

Gian Gastone Mascetti is a senior scholar of neurophysiology in the department of general psychology at the University of Padova in Italy, where he retired as a professor of neurophysiology. His research interests are focused on sleep, particularly the relation between sleep and brain lateralization.

Ultrasound Puts Animals into a Curious Hibernation-Like State

By Emily Willingham

In some species, when the going gets tough, the body hits the brakes, chilling body temperature and slowing metabolism to a snail's pace in a state known as torpor. Humans do not enter torpor, but the condition might offer benefits across scenarios as seemingly unrelated as intensive care unit (ICU) stays and long-distance space travel.

Using therapeutic ultrasound waves targeted at a deep region of the brain, researchers at Washington University in St. Louis and the University of Washington induced a torporlike state in mice without physically invading their skull. Mice can naturally enter this suspended state, so the scientists also tested the technique in rats, animals in which torpor is not part of their repertoire. The findings, published on May 25 in *Nature Metabolism*, could potentially inform research on targeting different brain areas with ultrasound to regulate other bodily activities.

The approach is noninvasive, which is "terrific for many applications," says Matteo Cerri, an associate professor at the department of biomedical and neuromotor sciences at the University of Bologna in Italy, who was not involved in the work. It seems to be flexible enough to act as a "thermostat, conceptually," allowing for adjustments of the ultrasound stimulus as needed to change temperature, he says.

Using a noninvasive technique is an important step in advancing torpor induction toward human use, says clinician Michael Ambler, a researcher and lecturer studying torpor at the School of Physiology, Pharmacology & Neuroscience at the University of Bristol in England, who was not involved in the work.

Some earlier efforts relied on injecting a gene for a protein into the targeted region to assist in activating cells under stimulation with light or drugs. Such invasive approaches are unlikely to gain approval for human use, Ambler says, making the new study "an extremely interesting piece of work."

To trigger a torporlike state noninvasively, researchers placed a tiny helmetlike probe on their rodent subjects that stimulated a deep brain structure with ultrasound waves. The team chose ultrasound frequencies that are different from those used for medical purposes, such as prenatal screening. The target region was the preoptic area of the hypothalamus, which contains neurons that previous work suggests play a role in torpor.

Stimulating these neurons prods them to send signals to brown adipose tissue, a highly metabolic fat located on the upper back that is tasked with turning up body heat when things get too chilled. The ultrasound-triggered messages from the preoptic area inhibited activity in the mice's brown fat, preventing it from elevating the heat. To confirm the chill, researchers used an infrared camera to track skin cooling over the brown fat area and heat loss in the animals' tail. They also confirmed the rodents' slowed metabolism by measuring the declines in their oxygen use. Along with the chilled fat and sluggish metabolism, the mice showed other signs of being in a torporlike state, such as reduced movement and a decreased heart rate.

Mice can enter into a natural torpor when frightened or stressed. To ensure that the ultrasound, not stress or fear, was triggering the temperature drop, the investigators turned to rats, which lack this natural response. The ultrasound signal also induced a colder body temperature in the rats, suggesting that ultrasound waves to the preoptic area were the cause of the torporlike state.

The effect in rats was mild, says Hong Chen, an associate professor at the department of biomedical engineering and radiation oncology at Washington University in St. Louis and senior author of the study. "The rat study was just proof of concept" that the stimulus would work in nontorporous animals, she says, adding that the research still has "a long way to go."

The investigators automated the stimulus process to keep the animals in a torporlike state. An uptick in temperature triggered the ultrasound stimulus, cooling things out again, much like a thermostat. Mice were kept in this state for 24 hours during the experiment, and when the ultrasound stimulation was switched off, normal temperature and metabolism was rapidly restored with no apparent negative aftermath.

A deeper look at what the cells were doing in response to the stimulus showed that the ultrasound waves affected the flow of ions such as calcium into the pre-optic neurons, triggering signals that reach the brown fat and keep it from warming things up. When investigators removed the protein responsible for controlling this flow, ultrasound had less effect on reducing body temperature.

These results imply that the protein is "like a nanoswitch," Chen says. She views this finding as the most important of the study because similar proteins in other brain regions also may be sensitive to the ultrasound stimulus. "If we can identify ultrasound-sensitive [proteins] in other parts of the brain, we may modulate other behaviors," she says, adding that what those might be remains to be seen.

The potential to alter behavior by bombarding the brain with ultrasound may have ethical implications. "In tech, we have to think about whether there's a dark side or not," Chen says. "I think the chance will be low because for this tech to work, we need well-designed devices that can precisely target a specific brain region, which is extremely challenging to do."

The technology will have to overcome initial skepticism. "As it stands, I think you can't really send someone into some degree of hypothermia" with a technique such as this, says Cerri, noting that there are other, far easier ways to render a person unconscious.

If noninvasive induction of torpor were to become possible in humans, one potential use would be to buy stroke or heart attack patients time in transport to the hospital, Chen says. Both of these emergencies result in oxygen deprivation to affected tissues, and because torpor involves a lower oxygen demand, the damage could be delayed or prevented.

In the ICU, torpor induction could preclude the need for the many drugs and monitoring involved in patient care. Ambler, who studies torpor as a way to support patients in the ICU who are undergoing organ failure, says, "This study presents a very first step toward that goal."

Torpor induction in humans also holds the futuristic potential to support suspended animation for journeys across vast stretches of lonely space.

Before that possibility takes flight in some aspirational future, research on a familiar earthbound species must come first. Cerri says the next step should be testing in larger, nonhuman animals, probably pigs. "They are most like humans–hairless and thermally similar to humans," he says.

Chen agrees that pigs probably are the next rung on the ladder to payoffs in torpor induction. "We want to push the technique step by step, from mice to rats to pigs to monkeys and then, hopefully, eventually to humans"–and then, perhaps, beyond the bounds of planet Earth.

About the Author

Emily Willingham is a science writer and author based in Texas. Her latest book is The Tailored Brain *(Basic Books, 2021).*

Sleeping with Half a Brain

By Christof Koch

> "The repose of the night does not belong to us. It is not the possession of our being. Sleep opens within us an inn for phantoms. In the morning we must sweep out the shadows."
>
> –Gaston Bachelard, French philosopher, 1960

Flies, birds, mice, dogs, monkeys and people all need to sleep. That is, they show daily periods of relative immobility and lack of response to external stimuli, such as light, sound or touch. This reduced sensitivity to external events distinguishes sleep from quiet resting, whereas the capacity to awaken from slumber distinguishes sleep from coma. Why sleep should be such a prominent feature of daily life across the animal kingdom, despite the fact that it leaves the sleeper unable to confront potential threats, remains mysterious.

Still, much progress in characterizing the physiology and capabilities of the sleeping brain has occurred over the past century, driven by the ability to record electrical activity of the brain (via electroencephalography, or EEG, on the surface of the skull), of the eyes (via electrooculography, or EOG), and of facial or other muscles (via electromyography, or EMG). For scientists, it is this triad of simultaneous measurements that operationally defines the state of sleep, leading to both surprising and counterintuitive insights.

Even without these tools, there are some basic things we do know about sleep. It is essential for our brain to function properly. Most of us have pulled all-nighters or have wanted to sleep but could not, unable to switch off our mind. The next day we are irritable, have trouble keeping our eyes open, and are terrible at tasks that demand sustained attention. Indeed, sleep deprivation causes many traffic accidents–the reason countries have laws that mandate a minimum rest period and maximum working hours for truck drivers.

Sleep is for the brain rather than for the body. Otherwise, eight hours of binge watching our favorite TV series in bed could replace sleep. Thus, we need to look at the sleeping brain to better understand the why and how of a state in which we pass one third of our life.

The Study of Shut-Eye

One milestone in the scientific study of sleep came in 1953, when Eugene Aserinsky and Nathaniel Kleitman of the University of Chicago discovered an until then unnoticed distinction in two distinct forms of sleep: rapid eye movement (REM) sleep and non-REM (NREM) sleep. When subjects are awake, before entering either of these states, their brain waves, recorded via EEG electrodes on the skull, display a typical pattern of electrical activity—low-amplitude, high-frequency signals—while their EMG reveals elevated muscle tone.

As individuals fall asleep and enter lighter and then deeper stages of NREM, also known as deep sleep, their brain waves progressively slow while increasing in amplitude. Eye movements, a hallmark of wakefulness, cease, and muscle tone diminishes. As sleep deepens, assayed by how difficult it is to wake a sleeper, so does the person's EEG. In the most restful form of sleep early on during the night, the EEG is dominated by high-amplitude waves, or oscillations, that slowly wax and wane. Electrical recordings of individual nerve cells in the neocortex directly underneath the skull show regular occurrences of on periods, when cells fire a series of all-or-none electrical pulses, called spikes, as happens when a person is awake. Pulses alternate with off periods, when neurons turn silent. These on and off periods and the associated slow waves in the EEG, termed slow-wave activity (SWA), occur as often as four times every second or as infrequently as once every four seconds (covering a frequency range from 0.25 to four hertz).

NREM sleep is interrupted by shorter episodes of REM sleep during which the EEG has a drastically different character: the slow and large waves are replaced by fast and choppy ones that superficially resemble the awake brain. The same paradoxical activation is seen at the level of individual neocortical neurons that fire spikes with the same intensity as they do during the day. Muscular tone is gone–to all intents and purposes the body is paralyzed–except for the breathing musculature and the jerky, rapid and symmetric movement in each eye that give this phase of sleep its name.

Most of the night is spent in NREM, with the most restorative deep sleep and its associated SWA, taking up 20 to 25 percent of a full night's slumber. Slow-wave activity is homeostatically regulated–that is, the longer somebody stays awake, the deeper and more frequent slow waves occur the following night. Conversely, early in the morning when sleep pressure has lessened, SWA diminishes, and sleep becomes shallower. Likewise, taking a nap reduces nighttime slow waves.

A number of consumer devices now on the market play regular soft tones through headphones at the same frequency as the SWA to entrain deep-sleep waves and thereby induce a more restful power sleep.

One Hemisphere on Watch

Until recently, deep sleep in humans was thought to be a global condition: a person is either asleep or awake but not both simultaneously. Put differently, their brain is either in deep sleep, as characterized by slow-wave activity, or awake, but not both. Yet birds and aquatic mammals such as dolphins and whales display the remarkable phenomenon of unihemispheric slow-wave sleep: one half of their brain is awake, including an open eye, and the other half shows the electrical signatures of sleep. This is most likely a protective mechanism, enabling the animal to fly or swim and monitor its environment for threats with one hemisphere while the other gets some rest.

It now turns out that even for people, there is more to sleep than meets the (shut) eye. Frequent travelers will be familiar with the first-night effect, the observation that the initial night in an unfamiliar place, whether a hotel, a friend's apartment or a tent, is less restful than subsequent nights. We find it more difficult to shut our mind down and wake up groggy. A team of researchers under Yuka Sasaki and Takeo Watanabe of Brown University set out to investigate this phenomenon

Eleven healthy volunteers slept for two nights inside an advanced neuroimaging scanner that allowed recordings of the brain's weak but ever changing magnetic field. The scientists focused on slow-wave activity, measuring its strength in four networks in the left cortical hemisphere and four in the right. Intriguingly, they found that the left cortical default-mode network, a set of interacting regions associated with mind wandering and daydreaming, had less SWA than the right one during the first night. No such imbalance emerged during a second night sleeping in the scanner. Also, the more asymmetric the pattern of SWA during the first night, the longer it took subjects to fall asleep. Part of the left hemisphere, in essence, is not sleeping as deeply as the right one during the first night.

To test the extent to which the left hemisphere is more vigilant in an unfamiliar environment, the team delivered tones via earphones to 13 subjects (different from the initial 11 volunteers, who were by now used to the sleep setup). Most of the tones were the same, but on rare occasions a different one sounded: beep, beep, beep, *boop*, beep, beep. The oddball tone drew attention and triggered a signature electrical response. When the deviant tone is played to the left ear, its output is predominantly relayed to the right cortical hemisphere, which shows the characteristic vigilance response.

During the first night, the left hemisphere had a more pronounced vigilance response to these deviant tones as compared with the right hemisphere. The enhanced vigilance response also led to the left brain being more frequently aroused (as defined by EEG criteria) than the right hemisphere.

During the second session in the scanner, both left and right hemispheres responded weakly and in the same manner to oddball sounds, as they did to the stereotyped beeps during both nights. If a brain network in the left cortical hemisphere acts as a night watchman for the sleeper, then an irregular event registered only by the left brain (via the right ear) should elicit a faster response than an oddball sound delivered to the right brain (via the left ear). This idea was tested in a third group of 11 volunteers: they had to lightly tap their fingers whenever they heard the sounds while asleep in the scanner. (I know, it doesn't sound like the most restful way to sleep; they were also not permitted to drink caffeine or alcohol or to take a daytime nap.) Sounds delivered to the ear projecting into the left hemisphere were much more likely to trigger an awakening during the first night than sounds to the opposite ear and hemisphere. This left-right asymmetry disappeared during the second night's sleep. Furthermore, it took the left brain less time to awaken in response to the deviant sound than the right brain.

In short, while sleeping in an unfamiliar place, the left cortical hemisphere is more vigilant and responds stronger and faster than the right one. Evolutionarily, this reaction makes a great deal of sense. It is important that a sentinel—here the left cortical default-mode network—monitors the unknown environment for threatening events while we sleep. The human brain, it turns out, is endowed with a less dramatic form of the unihemispheric sleep found in birds and some mammals. For humans, familiarity with a place breeds a deep night's sleep.

If we consider the individual we routinely share a bed with—whether spouse, partner or child—to be the most important social component of the environment, then I suspect that the left hemisphere might also be more watchful during the first night we sleep alone in our familiar bedroom. It knows something is amiss, and we'll sleep less restfully as a consequence.

About the Author

Christof Koch is chief scientist of MindScope at the Allen Institute in Seattle and of the Tiny Blue Dot Foundation in Santa Monica, as well as author of The Feeling of Life Itself—Why Consciousness Is Widespread but Can't Be Computed. *He serves on* Scientific American*'s board of advisers.*

Spiders Seem to Have REM-Like Sleep and May Even Dream

By Betsy Mason

Barred from her lab by pandemic restrictions, behavioral ecologist Daniela C. Rößler caught local jumping spiders and kept them in clear plastic boxes on her windowsill, planning to test their reactions to 3-D-printed models of predatory spiders. When she came home from dinner one night, though, she noticed something strange. "They were all hanging from the lids of their boxes," says Rößler, a postdoctoral researcher at the University of Konstanz in Germany. She had never seen jumping spiders suspended motionless on silk lines like this before. "I had no idea what happened," Rößler says. "I thought they were dead."

It turns out the jumping spiders were simply asleep—and that Rößler had discovered an alternate sleeping habit of the species *Evarcha arcuata*, which had been known to build silk sleeping dens in curled-up dead leaves. But the real surprise came when she decided to spy on them all night. She bought a cheap night-vision camera, duct-taped some magnifying lenses to it (*E. arcuata* is typically around six millimeters long) and aimed it at a sleeping female spider. What Rößler saw on the recording astonished her.

Mostly the spider just hung there. But then her legs started to twitch, and her abdomen and even her silk-producing spinnerets did so as well. Sometimes her legs curled in toward her sternum. With every spider Rößler recorded, these odd movements only appeared during distinct bouts that lasted a little more than a minute and occurred periodically throughout the night. "They were just uncontrollably twitching in a way that really looked a lot like when dogs or cats dream and have their little REM phases," she says.

Rapid eye movement (REM) sleep involves a state of partial or near-total muscle paralysis paired with an active, awakelike brain state, which is why it is sometimes called "paradoxical sleep." In humans, this state has been strongly linked with dreaming. Rößler and her colleagues wondered if the twitching spiders could be experiencing something like an REM phase of sleep and possibly even having dreams. "We were like, 'Okay, that would be insane,'" she says. Then she thought, "Let's figure it out," and immediately changed her research plans for the spiders.

There is abundant evidence for REM sleep or an analogous state in many mammals and birds. Scientists have also found something similar in two reptile species and hints of a state like it in zebra fish. Both octopuses and cuttlefish appear to have an REM phase, complete with eye movements, arm twitches, and rapid skin color and texture changes that resemble displays they perform when awake. Beyond those animals, there are glimmers but not much evidence of REM sleep in invertebrates, including insects and arachnids.

"It wouldn't surprise me at all if [jumping spiders] have dreams," says behavioral ecologist Lisa Taylor, who studies the spiders at the University of Florida and was not involved in the new research. "They live in such a rich sensory world, and we know they have amazing cognitive capabilities and memory."

The jumping spiders offered a unique opportunity to expand the realm of dreaming animals, in part because of particular aspects of their ocular anatomy. Though most spiders cannot move their eyes even while awake, jumping spiders have long tubes that shift their retinas around behind their large principal eyes. What's more, for the first several days of life, the spiders' exoskeleton is translucent, so those eye tubes are visible inside their head.

When Rößler recorded 34 sleeping spiderlings, she found that their twitches were accompanied by unmistakable eye-tube movements that did not happen during other phases of sleep. "It's beautiful. I mean, it's crazy. It immediately makes a sleep researcher think about rapid eye movement sleep," says entomologist Barrett Klein, who researches bee sleep at the University of Wisconsin–La Crosse and was not directly involved in Rößler's work. "And to have the first indication that you could study such a thing or that it's even relevant in arthropods is thrilling to me."

But it is too soon to say for sure that the spiders are experiencing something akin to REM sleep in humans, Klein cautions. The researchers first need to confirm the spiders are actually asleep during this phase by showing that they are less responsive to their environment, he says.

Rößler and her "dream team" of co-authors have already started those tests. And she points out that the leg curling is a particularly striking aspect of the spiders' REM-like phase because that pose is typically only seen in dead spiders. Spiders use hydraulic pressure maintained by muscles to keep their legs extended, and the curling could result from the muscle paralysis that typifies REM sleep. The team's initial findings were published on Monday in *Proceedings of the National Academy of Sciences USA*.

The videos alone are fairly convincing, says Niels Rattenborg, who studies bird sleep at the Max Planck Institute for Biological Intelligence in Seewiesen, Germany, and was not involved in the new study. "If I had to bet money, and this is not very scientific, but I will predict that they are asleep," in part because the movements do not look purposeful like the ones the spiders make when awake, he says.

To show the arachnids' sleep is REM-like, the scientists will also need evidence that the spiders' brains are active as they twitch and move their eyes, Rattenborg says. Measuring activity in a poppy-seed-sized brain will be challenging, but Rößler says there are potential ways to do it. For example, other scientists recently figured out how to insert an electrode into the brain of a different jumping spider species without deflating its pressurized body and killing it.

Studying REM sleep in a wide range of species, including these spiders, could help us better understand how it works and why it exists, Rattenborg says. Some have theorized that REM sleep's characteristic eye movements in humans reflect visual scenes playing out while dreaming. This raises the exciting possibility that other animals that have an REM-like state also experience dreams. Scientists of course cannot ask animals about their dreams, but measuring brain activity might someday be another way to get at the question. We are still learning which patterns of brain activity are correlated with dreaming in humans, says neuroethologist Teresa Iglesias, who studies cephalopod sleep at the Okinawa Institute of Science and Technology in Japan and was not involved in the jumping spider research. "So it is very early to expect we'll be able to identify dreaming in a physiological way in other animals," she says.

Yet as clues that nonhuman animals dream continue to accumulate, the philosophical implications are potentially huge, says David Peña-Guzmán, a philosopher at San Francisco State University and author of the new book *When Animals Dream: The Hidden World of Animal Consciousness.* Dreaming offers an entry point into questions of awareness in other animals: it is difficult to imagine that even a simple dream is possible without something like an ego or an "I" experiencing it, he adds. So if spiders dream, "it might mean that we start talking about spiders having something like a minimal self," says Peña-Guzmán, who was not involved with the spider research.

The jumping spiders' visible eye tubes might offer a way to test the theory that rapid eye movements are related to visual dream sequences and if those scenes are replays of things the arachnids witnessed while awake, Rößler says. It may be possible to play spiderlings a video of a simple scene of, say, a cricket hopping while tracking and measuring their eye movements and then see if those movements are re-created during sleep.

Rößler also wants to look for REM sleep in other spider species and points out that it might look very different in animals that rely more on senses other than vision, such as spiders that use vibrations in their webs to detect when prey is caught. "Maybe orb weavers dream in vibrations," she says. "I think the story is probably that REM is just as universal across the animal kingdom as sleep is, but we just haven't looked."

About the Author

Betsy Mason is a freelance journalist based in the San Francisco Bay Area and a 2022 Alicia Patterson Foundation fellow reporting on how the science of animal minds is changing how we think about us and them.

Cuttlefish Are Dazzling, But Do They Dream?

By Jim Daley

Cuttlefish are known for their sophisticated camouflage, as well as their kaleidoscopic displays for attracting mates and mesmerizing prey. These close relatives of squid and octopuses achieve such feats via millions of chromatophores–tiny sacs of pigment under the skin attached to muscles that squeeze or relax to push colors to the surface. In a new study, researchers report they have observed resting cuttlefish cyclically changing color and twitching their limbs in phases that resemble rapid eye movement (REM) sleep.

Many animals experience REM, a sleep phase that may involve eye movements and loss of muscle tone; in humans and some other mammals, it is often linked to dreaming. Study co-author Teresa Iglesias, a biologist at the Okinawa Institute of Science and Technology, cannot yet say whether cuttlefish experience the same kind of sleep that we do, but the animals' active phase during rest is similar to the REM sleep observed in humans and other vertebrates. Although the lineage that includes cuttlefish diverged from vertebrates around 500 million years ago, the study findings hint at a common evolutionary origin for sleep, she says.

Iglesias and her colleagues filmed cuttlefish in laboratory tanks for 24 to 48 hours at a time. While resting, the animals would demonstrate REM-like behavior for periods lasting between two and three minutes. They made sporadic arm and eye movements, and the chromatophores around their eyes got darker. Such phenomena were not observed during waking activity or inactive parts of their "sleep" cycles. The findings were published in January in the *Journal of Experimental Biology*.

"Sleep is a neural phenomenon more than a behavioral one," says psychologist Jennifer Mather, a cephalopod expert at the

University of Lethbridge in Alberta, who was not involved in the new research. Cycles of brain activity while an animal is quiescent, she explains, suggest sleep may play a role in stabilizing neural circuitry and processing newly acquired information. The REM-like behavior observed in cuttlefish could indicate a similar process is happening in their brain, she says.

But do cuttlefish dream? "We can speculate all we like," Mather says. "It's difficult because I don't [even] think we know why we dream." Iglesias says it is not impossible, but there is not enough evidence to answer this question. "For now," she says, "the biggest takeaway is that we need to dig deeper and keep an open mind."

About the Author

Jim Daley is a freelance journalist from Chicago. He writes about science and health.

GLOSSARY

circadian rhythm The 24-hour cycle the body—both physically and mentally—follows.

dementia A medical condition that causes a person to become confused or forgetful.

electroencephalogram (EEG) A medical test that records the electrical activity of the brain.

elusive Hard to find.

expedite To cause something to happen quickly.

functional magnetic resonance imaging (fMRI) A medical test that measures brain activity associated with blood flow.

hippocampus The part of the brain associated with learning and memory.

hypertension High blood pressure.

incapacitated To be unable to move or function.

malaise A general feeling of being unwell.

metabolic Having to do with the way a person, animal, or plant processes and uses food.

neurodegenerative The degeneration of the nervous system, sometimes the neurons of the brain.

nocturnal The state of being active at night.

obesity The condition of being significantly overweight.

pharmaceutical A drug or medicine.

replicate To copy.

restorative The ability to make a person feel strong or healthy again.

semilucid Partially clear or awake.

spontaneous Happening suddenly.

synapse A very small gap between two nerve cells.

FURTHER INFORMATION

"Brain Basics: Understanding Sleep," National Institute of Neurological Disorders and Stroke, https://www.ninds.nih.gov/health-information/public-education/brain-basics/brain-basics-understanding-sleep.

Bushey, Daniel, Giulio Tononi, and Chiara Cirelli. "Sleep and Synaptic Homeostasis: Structural Evidence in Drosophila," *Science,* June 24, 2011, https://www.science.org/doi/10.1126/science.1202839.

Castro, Jason. "The Brain May Disassemble Itself in Sleep," *Scientific American*, May 1, 2012, https://www.scientificamerican.com/article/sleeps-secret-repairs/.

Cirelli, Chiara, and Giulio Tononi. "Is Sleep Essential?" *PLOS Biology*, August 2008, https://journals.plos.org/plosbiology/article?id=10.1371/journal.pbio.0060216.

Diekelmann, Susanne, and Jan Born. "The Memory Function of Sleep," *Nature Reviews Neuroscience*, February 2010, https://www.nature.com/articles/nrn2762.

Paller, Ken A., and Delphine Oudiette. "You Really Can Learn in Your Sleep," *Scientific American*, November 1, 2018, https://www.scientificamerican.com/article/you-really-can-learn-in-your-sleep/.

Stickgold, Robert, and Jeffrey M. Ellenbogen. "Sleep on It: How Snoozing Makes You Smarter," *Scientific American*, August 1, 2008, https://www.scientificamerican.com/article/how-snoozing-makes-you-smarter/.

Vyazovskiy, Vladyslav V., and Umberto Olcese. "Local Sleep in Awake Rats," *Nature*, April 28, 2011, https://www.nature.com/articles/nature10009.

CITATIONS

Why Do We Sleep? by Teodora Stoica (January 25, 2019); 1.2 Spark Creativity with Thomas Edison's Napping Technique by Bret Stetka (December 9, 2021); 1.3 Deep Sleep Gives Your Brain a Deep Clean by Simon Makin (November 1, 2019); 1.4 New Hypothesis Explains Why We Sleep by Giulio Tononi and Chaira Cirelli (August 1, 2013); 1.5 Spiky 'Sleep Spindles' Linked to Acts of Learning by Rebecca Sohn (September 1, 2022); 1.6 What a Long-Ago Epidemic Teaches Us About Sleep by Christof Koch (March 1, 2016); 2.1 Why Do Some People Need Less Sleep? It's in Their DNA by Karen Weintraub (October 16, 2019); 2.2 Should We Make Daylight Saving Time Permanent? Let's Sleep on It by Randyn Charles Bartholomew (March 17, 2022); 2.3 The Secret to a Better Night's Sleep: A Sense of Purpose? by Daisy Grewal (October 18, 2017); 2.4 Good Snooze for Insomniacs–Therapy Is as Effective as Sleep Drugs by Jessica Schmerler (June 8, 2015); 2.5 Putting Insomnia on Ice by Joe Kloc (November 1, 2011); 3.1 The Sleep-Dementia Connection by Claudia Wallis (May 1, 2019); 3.2 Weekday-Weekend Sleep Imbalance Bad for Blood Sugar Regulation by Karen Hopkin (March 11, 2019); 3.3 Sleep's Role in Obesity, Schizophrenia, Diabetes ... Everything by Gary Stix (December 20, 2013); 3.4 Why Sleep-Deprived People Are More Selfish and Lonely by Daisy Yuhas (October 14, 2022); 3.5 Sleep Deprivation Shuts down Production of Essential Brain Proteins by Emily Willingham (October 10, 2019); 3.6 Acting Out Dreams Predicts Parkinson's and Other Brain Diseases by Diana Kwon (February 1, 2023); 3.7 Sleep Apnea is Different for Women by Claudia Wallis (April 1, 2020); 4.1 Why Rocking to Sleep Is a Matchless Sedative–and Elixir by Bret Stetka (January 24, 2019); 4.2 Let Teenagers Sleep by The Editors (February 1, 2023); 4.3 High School Students Need More Sleep and Later School Start Times by Thomas Franich, Christina Cardiel, Alia Carson, Marcos Martinez, Kian Mendoza, Sakari Morris, and Christian Ramirez Acabal (May 24, 2023); 4.4 A Study Power Move: Sleep Between Sessions by Steven C. Pan (October 25, 2016); 4.5 Sound Awake: 'Noisy' Neurons May Repeatedly Disrupt Your Sleep by Dana G. Smith (May 1, 2018); 4.6 Should You Give Your Kid Melatonin? by Elana Spivack (May 5, 2023); 5.1 Memory Wins When Dogs Sleep by Julie Hecht (February 7, 2017); 5.2 Meet the Animals That Literally Sleep with One Eye Open by Gian Gastone Mascetti (June 1, 2019); 5.3 Ultrasound Puts Animals into a Curious Hibernation-Like State by Emily Willingham (June 5, 2023); 5.4 Sleeping with Half a Brain by Christof Koch (September 1, 2016); 5.5 Spiders Seem to Have REM-Like Sleep and May Even Dream by Betsy Mason (August 8, 2022); 5.6 Cuttlefish Are Dazzling, But Do They Dream? by Jim Daley (May 1, 2019).

Each author biography was accurate at the time the article was originally published.

INDEX